SpringerBriefs in Electrical and Computer Engineering

SpringerBriefs in Speech Technology

Series Editor:
Amy Neustein

Editor's Note

The authors of this series have been hand selected. They comprise some of the most outstanding scientists—drawn from academia and private industry—whose research is marked by its novelty, applicability, and practicality in providing broad-based speech solutions. The Springer Briefs in Speech Technology series provides the latest findings in speech technology gleaned from comprehensive literature reviews and empirical investigations that are performed in both laboratory and real life settings. Some of the topics covered in this series include the presentation of real life commercial deployment of spoken dialog systems, contemporary methods of speech parameterization, developments in information security for automated speech, forensic speaker recognition, use of sophisticated speech analytics in call centers, and an exploration of new methods of soft computing for improving human–computer interaction. Those in academia, the private sector, the self service industry, law enforcement, and government intelligence are among the principal audience for this series, which is designed to serve as an important and essential reference guide for speech developers, system designers, speech engineers, linguists, and others. In particular, a major audience of readers will consist of researchers and technical experts in the automated call center industry where speech processing is a key component to the functioning of customer care contact centers.

Amy Neustein, Ph.D., serves as editor in chief of the International Journal of Speech Technology (Springer). She edited the recently published book Advances in Speech Recognition: Mobile Environments, Call Centers and Clinics (Springer 2010), and serves as quest columnist on speech processing for Womensenews. Dr. Neustein is the founder and CEO of Linguistic Technology Systems, a NJ-based think tank for intelligent design of advanced natural language-based emotion detection software to improve human response in monitoring recorded conversations of terror suspects and helpline calls.

Dr. Neustein's work appears in the peer review literature and in industry and mass media publications. Her academic books, which cover a range of political, social, and legal topics, have been cited in the Chronicles of Higher Education and have won her a pro Humanitate Literary Award. She serves on the visiting faculty of the National Judicial College and as a plenary speaker at conferences in artificial intelligence and computing. Dr. Neustein is a member of MIR (machine intelligence research) Labs, which does advanced work in computer technology to assist underdeveloped countries in improving their ability to cope with famine, disease/illness, and political and social affliction. She is a founding member of the New York City Speech Processing Consortium, a newly formed group of NY-based companies, publishing houses, and researchers dedicated to advancing speech technology research and development.

Manisha Kulshreshtha · Ramkumar Mathur

Dialect Accent Features for Establishing Speaker Identity

A Case Study

 Springer

Manisha Kulshreshtha
Haskins Laboratories
Yale University
New Haven, CT 06511, USA

Ramkumar Mathur
Department of Microbiology
and Immunology
Columbia University Medical Center
Columbia University
New York, NY 10032, USA

ISSN 2191-8112 e-ISSN 2191-8120
ISBN 978-1-4899-9939-9 ISBN 978-1-4614-1138-3 (eBook)
DOI 10.1007/978-1-4614-1138-3
Springer New York Dordrecht Heidelberg London

Printed on acid-free paper

Springer is part of Springer Science+Business Media (www.springer.com)

Preface

Voice identification is a very important tool for forensic investigators to resolve crime cases, where biometric features were examined to extract information about an individual. It is a widely established method similar to other biometric techniques, such as DNA profiling, fingerprinting, and handwriting characteristics, and used predominantly over the past decades. Information about speaker is extracted from recorded speech material called speaker profiling. Extracting information like sex, age, dialect background, and regional info underlies in the category of speaker profiling.

In this book, we discussed our study using dialects of Hindi language, widely spoken language in India. Khariboli, Bundeli, Kannauji, Haryanvi, Chattisgarhi, Marwari, and Bhojpuri dialects are chosen from different parts of the Hindi-speaking regions for the study. Twenty male and 20 female speakers were selected from each dialectal region keeping the selection criteria; a total number of 210 speakers were selected. Due to the close approximation to standard Hindi and the dialect that forms the basis of the modern standard Hindi, Khariboli is considered as the basic language. The prepared texts are transliterated using same script (Devnagri), but different vocabularies are used within the dialect group. Bhojpuri, Chattisgarhi, Kannauji, Marwari, Khariboli, Bundeli, and Haryanvi dialects are found to be unique for characterization in terms of vowel quality and vowel duration when compared with Khariboli.

In detail, the vowel quality and quantity of dialect speakers have been measured with the help of formant frequencies and vowel length and compared with Khariboli speakers. The spectrographic study of vowel quality and quantity for various dialects reveal that each dialect possesses its own vowel quality, and quantity, which is quite distiguishable when compared with Khariboli. Intonation and tone have also been observed and compared. Acoustic features associated with lexical tone and sentence intonation are also found unique and useful to dialect identification for speaker profiling. Essentially, this case study contributes to understand the speaker identification process, in a situation, where unknown speech sample is in different language/dialect from the recording of suspect. Our data establish that vowel quality, quantity, intonation, and tone of a speaker as compared with Khariboli (standard Hindi) could be the potential features for identification of dialect accent. Therefore, speaker identification is using the accent feature as the discriminating factor for forensic voice identification.

New Haven, CT, USA   Manisha Kulshreshtha

Acknowledgments

I would like to thank Dr. R. M. Sharma, Professor in the Department of Forensic Science, Patiala University, Punjab, India, and Dr. C. P. Singh, Assistant Director, Forensic Science Laboratory, New Delhi, India, for providing me a high infrastructure laboratory and supportive environment, ideal to generate material for the study. Thanks to Erica Davis for her sincere help. Special thanks to Dr. Amy Neustein, series editor of Springer Briefs in Speech Technology, for her generous and distinguished help.

Editor's Note

The authors of this series have been hand selected. They comprise some of the most outstanding scientists—drawn from academia and private industry—whose research is marked by its novelty, applicability, and practicality in providing broad-based speech solutions. The Springer Briefs in Speech Technology series provides the latest findings in speech technology gleaned from comprehensive literature reviews and empirical investigations that are performed in both laboratory and real life settings. Some of the topics covered in this series include the presentation of real life commercial deployment of spoken dialog systems, contemporary methods of speech parameterization, developments in information security for automated speech, forensic speaker recognition, use of sophisticated speech analytics in call centers, and an exploration of new methods of soft computing for improving human–computer interaction. Those in academia, the private sector, the self-service industry, law enforcement, and government intelligence are among the principal audience for this series, which is designed to serve as an important and essential reference guide for speech developers, system designers, speech engineers, linguists, and others. In particular, a major audience of readers will consist of researchers and technical experts in the automated call center industry where speech processing is a key component to the functioning of customer care contact centers.

Amy Neustein, Ph.D., serves as an editor in chief of the International Journal of Speech Technology (Springer). She edited the recently published book Advances in Speech Recognition: Mobile Environments, Call Centers and Clinics (Springer 2010) and serves as a quest columnist on speech processing for Womensenews. Dr. Neustein is the founder and CEO of Linguistic Technology Systems, a NJ-based think tank for intelligent design of advanced natural language-based emotion detection software to improve human response in monitoring recorded conversations of terror suspects and helpline calls.

Dr. Neustein's work appears in the peer review literature and in the industry and mass media publications. Her academic books, which cover a range of political, social, and legal topics, have been cited in the Chronicles of Higher Education and have won her a pro-Humanitate Literary Award. She serves on the visiting faculty of the National Judicial College and as a plenary speaker at conferences in artificial intelligence and computing. Dr. Neustein is a member of MIR (machine intelligence research) Labs, which does advanced work in computer technology to assist underdeveloped countries in improving their ability to cope with famine, disease/illness, and political and social affliction. She is a founding member of the New York City Speech Processing Consortium, a newly formed group of NY-based companies, publishing houses, and researchers dedicated to advancing speech technology research and development.

Contents

Chapter 1
Introduction

An important characteristic of speaker profiling is the dialectal accent feature which could ultimately establish the speaker's identity through his dialect. This feature also depends on educational background, mother tongue of the person, and the regional language. In fact, nonnative language is always influenced by the native language. Extracting information like sex, age, dialect background, and regional info underlies in the category of speaker profiling. Here, in this case study, Hindi, which is the official language of India, is chosen because of its various popular dialects. Khariboli, which is one of the chosen dialects, is considered as base language for comparison purposes because of its close approximation to standard Hindi. Moreover, it is the dialect that forms the basis of the modern standard Hindi.

Essentially, voice identification is comparison of original sample with the control sample and involves various steps before the investigator finally reaches an opinion. However, obtaining a control sample is not always easy as the suspect tends to disguise or use different dialect accents. In that situation, the comparison between these samples becomes tedious; therefore, the speaker profiling techniques are used when you need to know the information about the speaker or suspect.

Voice identification is one of the commonly used methods by forensic experts to establish the identity of a person if the available evidences are in the form of a recorded conversation. It is also well established and widely accepted in jurisdiction decisions as a strong court evidences. Typically, it benefits to unravel a variety of criminal cases, including murder, rape, extortion, drug smuggling, wagering-gambling investigations, political corruption, money-laundering, tax evasion, burglary, bomb threats, terrorist activities, and organized crime activities.

The pioneer work for visual representations of speech sound was developed using a spectrograph, an automatic sound wave analyzer that generates the visual prints of a speech in terms of frequency, intensity, and time [32]. A spectrograph was also widely used during World War II, where with the help of spectrograms enemy voices were identified more easily. In 1962, Kersta et al. [39] found that the spectrograms (named as voiceprints) obtained through the spectrograph were unique and individualistic in nature. He found that a person's "voiceprints" were permanent

M. Kulshreshtha and R. Mathur, *Dialect Accent Features for Establishing Speaker Identity: A Case Study*, SpringerBriefs in Electrical and Computer Engineering, DOI 10.1007/978-1-4614-1138-3_1, © Manisha Kulshreshtha 2012

and remained unchanged throughout the lifetime of that person, even as the person grows old, loses tonsils, teeth, or adenoids. Kersta stressed that this is true even when voice disguise or mimicry has no effect on the voiceprints.

Bolt et al. published a paper (1969 and in 1970) indicating the necessity of performing an experimental study in order to determine the reliability of identification or elimination of an unknown speaker among the known ones by examining their speech spectrograms [9]. In that study, the authors suggested that the testing of several variables correlated with forensic models of identification of talkers, since these variables have not been studied previously and because of the interest of determining the feasibility of applying this method of identification as a means to obtain legal evidence. A large experimental study that included most of the forensic variables was initiated by Tosi and associated at Michigan State University early in 1968 and concluded in 1970 [56].

The number of variables involved in comparing voiceprint evidence with that of handwriting identification evidence is equally large. Handwriting is also amenable to disguise and forgery and suffers deterioration due to disease, old age, injury, and influence of intoxication. However, a lot of inputs are required in case of handwriting evaluation and it has achieved a high degree of reliability. Similar extensive inputs were needed before the voiceprint identification could achieve a reasonably high degree of reliability. If voice identification does not provide positive proof by itself, it can certainly provide corroboration, indicate leads, and help investigations in other ways. While there is disagreement in the scientific community on the degree of accuracy with which examiners can identify speakers under all conditions, there is agreement that voices can, in fact, be identified.

1.1   Voice Identification Theory

Theory of voice identification is based on the fact that every voice has individual characteristic features to distinguish it from others. Every voice is unique because each individual has different size of the vocal cavities (throat, nasal, and oral cavities), shape, length, and tension of the individual's vocal cords located in the larynx. Principally, the vocal cavities behave as resonators, much like organ pipes, which reinforce some of the overtones, produced by the vocal cords and produce frequency formats or voiceprint bars. Thus, the possibility of two people having similar sizes of vocal cavities and similar oscillations seems very rare. Another important factor is the manner in which the articulators or muscles of speech are manipulated during speech. The articulators are lips, teeth, tongue, soft palate, and jaw muscles which help in producing speech by controlling the muscles. The possibility of two people developing identical manner of using articulators also appears very remote. Therefore, there is a very small likelihood that two speakers would have identical vocal cavity dimensions and configurations with identical articulator use patterns.

In general, the voice identification comprises both aural and visual (spectrographic) methods for identification purposes. In typical voice identification case, the examiner is usually given various recording samples; one is questioned sample from which the voice is to be identified; another one is called specimen sample or control sample, which is the recording of one or more suspects. The examiner has to correctly examine the identity of the known voice from these recordings. In order to evaluate the recording of the unknown voice, the very first step is to examine the clarity and sufficiency of contents, and also whether the sample is in the required frequency range for analysis or not. In addition, the volume of the recorded voice signal must be significantly higher than that of the environmental noise. If the noise, music, and voices of other speakers are large, then the recorded speech samples must be longer. Even, if the quality of recordings is not good enough, the examiners can exclude the irrelevant contents. It is always very easy to compare speech samples that are the same as the text of the recording in questioned sample. Generally, the suspects repeat the text of the recording of the unknown speaker several times and these words will be recorded in a similar manner to the recording of the unknown speaker, that way the examiner can have similar speech sound for comparison as well as the information about the sound transition. In a situation, where voice sample must be obtained without the knowledge of the suspect, like interception, the amount of speech necessary to do the comparison is usually much greater. In order to get the required voice recording when the suspect is in a conversation, the conversation must be manipulated in such a way that the suspect repeats as many of the words and phrases found in the text of the unknown recording as possible. It is always recommended to record the large number of data to have sufficient amount of comparable speech. Once the sufficient evidence has been collected to perform the analysis, the two-step process of voice sample comparison is used. One is aural analysis (listening) and the other is spectrographic analysis (visual). Though, these are two separate steps but have equal importance and the examiner combines them to reach the final conclusion. In an aural comparison of the voice samples, the examiner compares single speech sounds as well as series of speech sounds of the known and unknown samples. The examiner also conducts a number of tasks like comparing samples to obtain similarities and differences, screening out less useful portions of the samples, and indexing the samples for further analysis. One of the most commonly used methods of aural comparison is re-recording of a speech sound sample of the unknown and followed by immediately re-recording of the same speech sounds of the suspect. This is repeated several times so that the final product is a recording of specific speech sounds in alternating order, first by the unknown speaker, then, by the suspect. During the aural comparison, the examiner studies the psycholinguistic features of the speakers' voice. There are a large number of qualities that are examined from accent and dialect, syllable grouping and breath patterns. The examiner also looks for signs of speech pathologies and peculiar speech habits.

The second step in the voice identification process is spectrographic analysis of the recorded samples. Sound spectrograph is an automatic sound wave analyzer with a high-quality, fully functional tape recorder. The speech samples to be analyzed are recorded on the sound spectrograph analyzed in a form of spectrogram. The spectrogram is a graphic display of the recorded signal on the basis of time and frequency with a general indication of amplitude. The spectrograms of the unknown

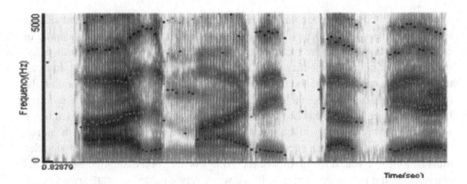

Fig. 1.1 Visual representation of speech (spectrogram)

speaker are then visually compared with the spectrograms of the suspects and the speech sounds used for comparison. The comparisons of the spectrograms are based on the acoustical features representing in terms of displayed patterns of the captured speech. The examiner studies the bandwidths, mean frequencies, and trajectory of vowel formants; vertical striations, distribution of formant energy, and nasal resonances; stops, plosives, and fricatives; inter formant features, the relation of all features present as affected during articulatory changes, and any peculiar acoustic patterning. The speech samples are closely examined to look for similarities and differences, if they are due to pronunciation differences or if they are indicative of different speakers. When the analysis is complete, the findings from both the aural and spectrographic analyses are combined into one of five standard conclusions: a positive identification, a probable identification, a positive elimination, a probable elimination, or no decision. For a positive identification, the examiner must find a minimum of 20 speech sounds that possess sufficient aural and spectrographic similarities. The probable identification conclusion is reached when there are less than 20 similarities and no unexplained differences. This conclusion is usually reached when working with small samples, random speech samples, or recordings of lower quality. The result of positive elimination is rendered when 20 differences between the samples are found that cannot be based on any fact other than different voices having produced the samples. A probable elimination decision is usually reached when working with limited text or a recording of lower quality. The no decision conclusion is used when the quality of the recording is so poor that there is insufficient information with which to work.

The spectrogram displays the speech signal with the time along the horizontal axis, frequency on the vertical axis, and relative amplitude indicated by the degree of gray shading in the display (Fig. 1.1). The resonance of the speaker's voice is displayed in the form of vertical signal impressions or markings for consonant sounds, and horizontal bars or formants for vowel sounds. The visible configurations are characteristic of the articulation involved for the speaker producing the words and phrases. The spectrograms serve as a permanent record of the words spoken and facilitate the visual comparison of similar words spoken between and unknown and known speaker's voice.

The visual displays of the disputed and the sample utterance (of the same word or text) are compared visually. The process of using the method is simple. The suspects are made to utter relevant word(s) and the voice is recorded on a device similar to one used in recording the disputed utterance. The recorded voices are then turned into their visual spectrograms through the sound spectrograph and the spectrograms are compared and evaluated. The resonance and the reverberation of the room, where the suspect is speaking through telephone, may cause damping or amplification by some bands of frequency. The unknown and known samples are to be recorded within the same room in order to minimize resonance effects on the process of speaker identification. In actual cases, the environment in which the criminal is using the telephone is not known. However, it is demonstrated by other group where the effect of this resonance is not significant in most of the cases [56].

Another factor which could affect the recorded speech is due to the resonance or response curve of the telephone line utilized. The response curve of the telephone line may amplify or attenuate some band of frequency within the range of frequency, normally from 150 to 4,000 Hz. The pickup device and method of connecting the receiving telephone to the recorder and also the tape recorder itself can also introduce distortion to the unknown and known samples. In every case the response curve of each of the transmitting and recording elements in use during the recording of both known and unknown voice should be obtained to have information of this type of information [56]. On the other hand, speech sounds are generally posed with variability during the production of speech. Generally speaking, variability in speech can be of two types, i.e., intra-speaker and inter-speaker variability. Intra-speaker variability, which exists within the same speaker, is due to many factors. Some of these factors are emotions, rate of utterance, mode of speech, disease, mood of the speaker, and the emphasis given to a word at a particular moment. One of the important factors without measurements is the temporal variation within the speaker. The inter-speaker variation, which exist among the different speakers, arises mainly due to anatomical differences in the vocal organs and from learned differences in the use of speech mechanism.

1.2   Tests and Errors in Voice Identification

According to the composition of unknown and known voice samples, tests of voice identification or elimination can be classified into three group; discrimination tests, open tests, and close tests. In the discrimination tests, the examiner is provided with one unknown voice sample and one known voice sample. Two types of errors can be produced in the discrimination tests: (a) false elimination, when the examiner decides that both samples belong to different talkers, but they are actually from the same talkers, and (b) false identification, when the examiner decides both samples belong to the same talkers, when they actually do not. In the open tests, the examiner is given one unknown voice sample and several known voice samples. He is told that the unknown sample may or may not be found among the known samples. This type of test can yield three types of errors. The first is false elimination, when the unknown voice sample is among the known samples. But the examiner decides that it is not. The second and third types are errors of false identification that can originate from two possibilities (a) when one of the known samples is the same as the unknown one, and (b) none of the known samples is same as the unknown one and the examiner decides that one of them is same as the unknown. In the closed tests of voice identification, the examiner is given one unknown voice sample and several known voice samples, but he is told that the unknown voice sample is also included in the known voice samples. Consequently, here only one type of error may be produced, i.e., an error of false identification in which the examiner selects the wrong one.

1.3 Speech Production Mechanism

Speech sounds are produced by the movements of the various vocal organs, interrupting the current of air movement from and into the lungs. Speech production, basically, is a combination of air steam and process of articulation, which is discussed in the following part of the chapter. The air column from larynx to the lips consists of four cavities, namely, the oral cavity, the nasal cavity, the pharyngeal cavity, and the lungs with interconnecting passages at the back of the mouth (Pharynx) and the glottis as shown in Fig. 1.2. The movement of jaw, lips, tongue, soft palate, wall of the pharynx, and vocal folds to alter the shape of vocal tract regulate the air stream from lungs to atmosphere. An air stream initiated by the lungs is called the pulmonic air stream. Most of the human speech sounds have the pulmonic exgressive air stream as their source.

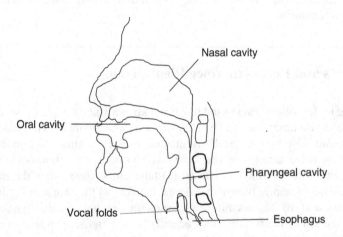

Fig. 1.2 Four major vocal cavities of vocal tract

1.4   Articulators

Parts of the vocal tract that can be used to produce distinctive sounds are called articulators. Articulators can be grouped into active and passive articulators on the basis of their activity. The articulators that move during the process of articulation are called *active articulators*, whereas organs of speech, which remain relatively motionless, are called *passive articulators*.

Active articulators include the lips (lower and upper both), glottis, vocal folds, and tongue. For the sake of convenience, tongue is divided into different parts, viz., the tip, the blade, the front, the back, and the root of the tongue. When the tongue is at rest behind the lower teeth, then the front part of the tongue, which lies opposite to the hard palate, is called the front of the tongue. The part which faces the soft palate is called the back, and the region where the front and back meet is known as center. Some scholars call the whole upper surface of the tongue, i.e., the part lying below the hard and soft palate, as the dorsum. The part of the tongue at the extreme end and facing the tongue can be further divided into three parts as follows; the tapering section of the front facing the teeth ridge is called the blade. Its extremity is called the tip. The tip and the blade of the tongue are together called sometimes the apex. The remaining part of the front of the tongue is referred to as front only.

Passive articulators include the teeth, just behind the upper teeth there is a small protuberance called alveolar ridge, the front part the roof of the mouth is formed by a bony structure known as hard palate. The hard palate is immediately followed by the soft palate or the velum. It is like a soft muscular sheet attached to the hard palate at one end, and ending in a pendulum like soft muscular projection at the other which is called the uvula and epiglottis. Most of the active articulators are attached to the movable lower jaw and as such lie on the lower side or the floor of the mouth. Passive articulators, attached to the immovable upper jaw, lie on the roof of the mouth. Figure 1.3 clearly shows all the active and passive articulators involved in speech production.

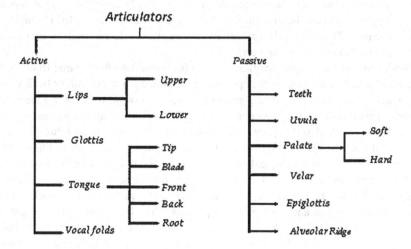

Fig. 1.3 Various articulators

The points at which the articulators are moving toward or coming into contact with certain other organ are called *place of articulation*. The type or the nature of movement made by the articulator is called the *manner of articulation*. Therefore, nearly all the articulatory description of a speech sound has to take into consideration the articulator, the point/place of articulation, and the manner of articulation.

1.5   Articulation

The distinction between vowels and consonants can be done on the basis of manner of articulation of speech sound. In the production of vowels, the outgoing air stream passes freely through the oral cavity that acts as a resonator, while in the production of consonants the air stream meets some type of obstructions before it finally passes out of the cavity.

1.5.1   Articulation of Vowels

A vowel is a voiced sound that can be produced in isolation without changing the position of articulators using glottis as a primary source of sounds with no friction of air against the vocal tract. Vowels are produced in a high, middle, or low position of the tongue. In the production of vowel sounds, none of the articulators come very close together and the passage of the air stream is relatively unobstructed. Vowel sounds are specified in terms of the position of the highest point of the tongue and the position of the lips. Generally, four vertical levels are recognized as reference points.

The two extremes as the highest and lowest possible levels are termed as "high" and "low" and intermediate level are named as "higher mid" and "lower mid." Alternative terms used for these positions are (1) close, (2) half-close, (3) half-open, and (4) open. The vowels, thus produced in these positions, are called (1) high or close vowels, (2) higher mid or half-close vowels, (3) lower mid or half-open vowels, and (4) low or open vowels.

With reference to tongue position, a plane is divided into three parts: front, central, and back, and the vowels thus produced are called front, central, or back vowels. These positions of the vowels are represented in a two-dimensional schematic diagram called vowel quadrilateral as shown in Fig. 1.4, for example, vowels in /tin/ (three), /din/ (day), /dɛr/ (late), /pɛr/ (feet), /kɑm/ (work), /tum/ (you), /pʰul/ (flower). In the first four vowels, the highest point of the tongue is in the front of the mouth. Accordingly, these vowels are called front vowels. The tongue is fairly close to the roof of the mouth for the vowel in / tin /, slightly less close for the vowel in / din /, and lower still for the vowels in / dɛr / and / pɛr /. The vowel in / tin / is classified as a high front vowel and the vowel in / pɛr / as a low front vowel. The height of the tongue for the vowels in the other words is between these two extremes, and they are therefore called mid front vowels. The vowel in / din / is a high mid vowel and the vowel in / dɛr / is a low mid vowel.

Fig. 1.4 The vowel
quadrilateral

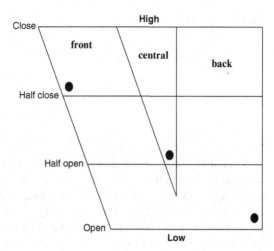

In other three vowels, the tongue is close to the upper or back surface of the vocal tract. These vowels are classified as back vowels. The body of the tongue is highest for the vowel in / pʰul /, which is therefore called a high back vowel and lowest for the first vowel in / kɑm /, which is therefore called a low back vowel. The vowel in / tum / is a mid-back vowel.

The position of the lips also varies considerably in different vowels. In the last two words, there is a movement of the lips in addition to the movement that occurs because of the lowering and rising of the jaw. This movement is called lip rounding. Based on the movement of the lips, vowels are described as rounded or unrounded vowels.

1.5.2   Articulation of Consonants

Consonants are phonemes other than vowels or glides. Consonants always include friction and they can be classified into voiced and voiceless, also on the basis of the manner of articulation and place of articulation as stop, nasals, fricatives, trills, flaps, laterals, affricates, continuants, etc. During the process of articulation, an articulator moves toward or comes into contact with a point of articulation in such a way that the passing air stream is obstructed or modified so as to produce some type of audible noise. Different manners of articulation produce different types of sounds.

1.6   Prosodic Features

Suprasegmental or prosodic features are often used in the context of speech to make it more meaningful and effective. Without prosodic features superimposed on the segmental features, a continuous speech can also convey meaning but often loses the effectiveness of the message being conveyed. The most widely discussed phonological suprasegmental is the syllable.

Fig. 1.5 Parts of a
monosyllabic word

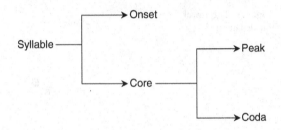

A **syllable** is considered as a phonological unit and generally consists of three phonetic parts, namely, *the onset*, *the peak*, and *the coda* (Fig. 1.5). In a monosyllabic word /bat/, /b/ is the onset, /a/ is the peak, and, /t/ is the coda. A syllable is further discriminated as closed syllable comprising a vowel (V) as a nuclei, preceded and followed by consonants (C) as onset and coda, i.e., CVC or VCV structure and open syllable, in which either onset or coda is absent (CV or VC). Features like stress, tone, and duration are always present in almost all the utterances of a language. Thus, all the utterances can be characterized by different degrees of these features.

In terms of linguistic, prosody is rhythm, stress, and intonation of speech. Basically, prosody corresponds to various features of the speaker such as the emotional condition; the form in which the sentence or word is uttered, i.e., question, command, or statement. It also reflects the presence of emphasis, contrast, and focus in the utterance. In terms of acoustics, the prosodic of a language involves variation in formant frequencies, pitch, loudness, and the syllable length of speech sounds. The details of a language's prosody depend on its phonology. The vowels and consonants are considered as small segments of the speech, which together form a syllable and make the utterance. Specific features that are superimposed on the utterance of the speech are known as suprasegmental features or prosodic features. Common prosodic features are the *stress, tone,* and *duration* in the syllable or word for a continuous speech sequence. Sometimes even harmony and nasalization are also included under this category.

1.6.1 Stress

Stress is one of the prosodic features of utterances. It applies not to individual vowels and consonants but to the whole syllable. A greater amount of energy is required to pronounce a stressed syllable than that of the unstressed. Since more air is pushed out from the lungs by extra contraction of the abdominal muscles in addition to the laryngeal muscles, there is an additional rise in pitch. An increase in the amount of air pushed out also increases the loudness of the sound produced.

There is a misconception of expressing stress in terms of loudness; loudness is simply the amount of acoustic energy involved in the process. There are many more factors affecting the stress like length, change of pitch, and quality of the vowel. These features all together can express the variation of stress over utterances in a language.

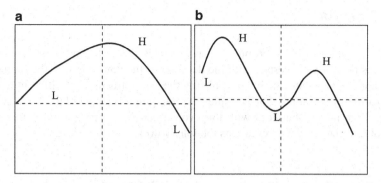

Fig. 1.6 Pitch variation in two different languages

In a situation where words involve more than one syllable, a language-specific stress pattern is expected. Disyllabic words are distinguished as stressed and unstressed, whereas a polysyllabic word is distinguished as primary stressed/unstressed and secondary stressed/unstressed.

1.6.2 Intonation

Intonation can be defined acoustically with respect to time as the gradual variation of fundamental frequency. If the variation is along the sentence or phrase, it is termed as *sentence/phrasal intonation* whereas along the word/word-segment it is called *lexical intonation* or tone and the languages with such features are known as tonal languages. More analytically, intonation is something which is superimposed upon the meaning of words uttered and carries important semantics function in many languages. It is important to note that such semantic functions through intonations can be brought out in the speech continuum by rhythmic modulations.

1.6.3 Tone

Tone may be rising, level, or falling tone depending on the increase, steady, and decreasing of pitch, respectively. Tone or intonation of a speaker can be represented as low (L), high (H), or the combination of these two. The intonation pattern in case of two different accents of two languages is shown in Fig. 1.6. Rules of representation, which are common to all the languages, are known as *conventional rules,* whereas, dialect/language-specific rules are termed as *prescription rules*.

1.6.4   Length

Languages can also be discriminated on the basis of the length of the segment. Length is basically concerned with the duration of the vowel used in the segment. There are languages in which words having the same structure but different length convey different meaning. For example in Hindi /gɑnɑ/ is the word used for "song" whereas, /gɑ:nɑ / is the word with long vowel conveying the meaning "to sing." Symbol /:/ after the vowel represents the long vowel.

1.7   Acoustic Characteristics of Speech Signals

The study of sound is called acoustics. Most of the sounds we hear such as the street noises, the sounds of speech, and even the sounds of music are complex as they consist of many tones or frequencies. A pure tone has only one frequency of vibration. It is a result of vibrations that repeats itself at a constant number of cycles per second. The number of cycles per second is termed as its *frequency*.

In the course of any speech sequence, the fundamental frequency produced by the larynx changes continuously and hence, as we have seen, the series of harmonic frequencies also changes. It is for this reason that the formants play such an important role in speech reception, since they give rise to peaks of energy in the spectrum, which are relatively independent of the fundamental frequency. It follows that in acoustic study of speech we are not so much interested in what is happening to the harmonics of the larynx tone as in what is happening to the formants of the vocal tract.

1.8   Frequency and Pitch

People differ in the frequency range to which their ears are tuned, but in general, young, healthy, human ears can detect vibrations as low as 20 Hz and as high as 20,000 Hz. Vibrations too low in frequency to be audible are called *subsonic*, and those too high *ultrasonic*. Extremely low frequency sounds may not be perceived by human ear, but it is often felt. The frequencies important to the speech signal are within the 100–5,000 Hz range.

Frequency relates directly to pitch, which is a sensation to listener. In general, when frequency of vibration is increased, a rise in pitch is perceived and when frequency is decreased, a lowering of pitch is perceived. Pitch, in contrast, is a psychological phenomenon. It is the way in which frequency the listener perceives changes and differences. In the frequencies below 1,000 Hz, perceived pitch is fairly linear in its relationship to frequency but as the frequencies get higher, it takes a larger change in frequency to effect a change in the sensation of pitch. The relationship between the physical property of frequency and psychological sensation of pitch is illustrated in Fig. 1.7.

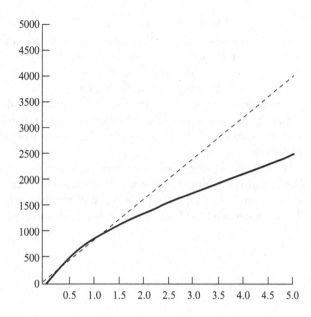

Fig. 1.7 Relationship between frequency and pitch

1.9 Intensity and Loudness

The amplitude of vibration and the extent of particle displacement indicate the intensity of sound. A unit of measurement called a decibel (dB) is used to describe the relative intensity of two sounds.

The decibel (dB) is used to measure sound level, but it is widely used in electronics, signals, and communication. The dB is a logarithmic unit used to describe a ratio. The ratio may be power, sound pressure, voltage or intensity, or several other things. Later on we relate dB to the "phon" and "sone" (other units related to loudness) [28]. The name Bell was given to commemorate the work of Alexander Graham Bell, the inventor of the telephone.

1.10 Speech Digitization

By treating the speech signal as if it is made up of periodic functions, the principles of sampling and quantization can be applied for digitizing speech samples. Based on Fourier analysis, the speech signal is comprised of sinusoidal waves of varying frequencies. To represent the speech signal, a sampling rate greater than twice the highest frequency of any sinusoidal component is required.

The sampling rate is the number of points looked at each second. A sampling rate of 10,000 would record values 10,000 times each second, or once in a 0.1 millisecond (ms). The sampling period is the amount of time that elapses between each sample. The sampling period is the inverse of the sampling rate and vice versa. If the amplitude of the signal is unknown, the sampling rate must be larger than twice the frequency. This avoids the possibility of having all samples on zero crossing, which would allow the reconstruction of sine waves of a single frequency, but with many different types of amplitude.

A sampling rate of twice the signal frequency is called the *Nyquist rate*. The sampling theorem gives us the minimum sampling rate for the accurate reconstruction of a sine wave. The circumstances that results from the under sampling rate is called *aliasing*.

Quantization is the process of representing a range of numbers with a fixed number of digits, rounding to the nearest representable necessary value. In computers, it is the process of representing a range of numbers with a fixed number of bits. The device that performs quantization is called quantizer.

Chapter 2
Hindi Language and Its Dialects

In Chap. 1, we discussed the basics of voice identification process, which is very useful to forensic expert, in order to correctly identify the speaker. In this study, we are trying to establish the identity of a person on the basis of his/her dialectal accent. Therefore, it is recommended to have a basic knowledge of the native language and its dialects.

Language is an arbitrary system by which one can exchange his or her thoughts and ideas with others. It has a set of specified symbols (graphemes) assigned to specific phones called a script of language. The script contains the words or vocabulary used within a linguistic community to express their views as well as to convey the meanings according to the prescribed rules. Different communities are using different sets of sounds for communication and, thus, have different languages. Versions of a language that sound different but are mutually intelligible are called dialects of the language. Generally, dialects of a language do not have any written script. Speaker, while speaking different dialect, uses same basic language but different vocabularies and also pronounces words differently. Because of this, speech of a person belonging to a dialectal group always has its specific dialectal accent. In fact, native accent is always affected while speaking nonnative language and shows the potential in the investigation of crime as accented speech carries linguistic information regarding the regional dialect of an individual. In this case study, we choose Hindi language of India and studied some of its popular dialect to observe the effect of dialect accent on person's speech and to use these features for identification and profiling purposes.

2.1 Hindi Language

Hindi is the official language of India and it is spoken by the majority of the population of India. This is the reason I chose to select the Hindi language and its dialect for the study. Hindi is one of the Middle Indo-Aryan languages. There is no specific time mentioned in the literature about the birth of Hindi, but 1000 AD is commonly

M. Kulshreshtha and R. Mathur, *Dialect Accent Features for Establishing Speaker Identity:* 15
A Case Study, SpringerBriefs in Electrical and Computer Engineering,
DOI 10.1007/978-1-4614-1138-3_2, © Manisha Kulshreshtha 2012

accepted. Hindi sometimes is also referred as Hindavi or Hindustani and also Khariboli. Standard Hindi, or more precisely Modern Standard Hindi, also known as Manak Hindi (Devnagri: मानक हिन्दी), High Hindi, Nagari Hindi, and Literary Hindi. It is a standardized and sanskritized register of the Hindustani language derived from the Khariboli dialect of Delhi (and the surrounding western Uttar Pradesh and southern Uttarakhand region). Hindi assumes much importance as it is spoken by a large number of people across the globe. The apbransh of Prakrit and Pali led to the rise of different regional dialects of Hindi. These dialects paved the path for Khariboli, a form in which Hindi is recognized today. Over nearly a thousand years of Muslim influence, when Muslim rulers controlled much of northern India during the Mughal Empire, many Persian and Arabic words were borrowed [34].

Modern Hindi is mutually intelligible with the alternative register of the Urdu language. Mutual intelligibility decreases in literary and specialized contexts which rely on educated vocabulary. Because of religious nationalism since the partition of British India and continued communal tensions, native speakers of both Hindi and Urdu frequently assert them to be completely distinct languages, despite the fact that they generally cannot tell the colloquial languages apart.

The dialect upon which Standard Hindi is based is Khariboli, the vernacular of the Delhi and Uttar Pradesh region. This dialect acquired linguistic in the Mughal Empire (seventeenth century) and became known as Urdu, "the language of the court." After independence, the Government of India set about standardizing Hindi as a separate language from Urdu.

2.2   Varieties (Dialects) of Hindi

Hindi (in the broad sense) is a subset of the Indo-Aryan language family in the northern plains of India, bounded on the northwest and west by Punjabi, Sindhi, Gujarati, and Marathi; on the east by Maithili and Bengali; and on the north by Nepali. Hindi covers a number of Central, East-Central, Eastern, and Northern Zone languages, including the Bihari languages excepting Maithili, the Rajasthani languages, and the Pahari languages excepting Dogri and Nepali.

Hindi is often divided into western Hindi and eastern Hindi, which are further divided into its various dialects. Around 200 regional dialects are identified within Hindi language itself. Many linguists consider only western and eastern dialects as the proper dialects of Hindi. The rest are considered as sublanguage or separate language. There is a report [56] noted that the classification of the dialect under various branches and their classification as a dialect of Hindi or as an independent language depend on the perception of the linguist. Accordingly, dialects of Hindi are categorized into five groups and further into subgroups listed in Fig. 2.1. For the purpose of experiments, some popular dialects of Hindi, namely, Khariboli, Bundeli, Kannauji, and Haryanvi among the western Hindi belt, Chattisgarhi from eastern Hindi dialectal group, Marwari from Rajasthani group, and Bhojpuri from Bihari group of Hindi belt, have been selected.

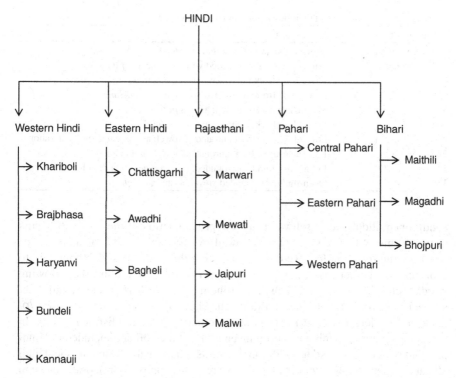

Fig. 2.1 Various dialects of Hindi language

2.3 Phonology of Hindi Language

Phonology is a subfield of linguistics, which studies the sound system of languages. An important part of phonology is to study distinctive sounds within a language. There are 11 vowels and 35 consonants frequently used in Hindi speech with a few sounds borrowed from Persian and Arabic languages but now considered as a part of Hindi language. These sounds can be further classified according to the place and manner of the articulation. Vowels of Hindi are almost similar to that of other languages but there are certain differences and special features in Hindi consonants, which are common to other languages.

2.3.1 Vowels of Hindi

Vowels of Hindi are classified into 11 different vowel sounds as mentioned in various literatures. In addition, there are extra sounds, which sometimes are included as a part of vowel sound system of Hindi language. Vowel sound system of Hindi language is described in Table 2.1. The short-open-mid-unrounded vowel /ɛ/ (as e in get) does not have any symbol or diacritic in Hindi script. It occurs only as

Table 2.1 Vowels sounds and their description in Hindi language

Alphabet	Pronunciation	Description
अ	/ə/	Short or long Schwa: as the a in *above* or *ago*
आ	/a:/	Long open back unrounded vowel: as the a in *father*
इ	/i/	Short close front unrounded vowel: as i in *bit*
ई	/i:/	Long close front unrounded vowel: as i in *machine*
उ	/u/	Short close back rounded: as u in *put*
ऊ	/u:/	Long close back rounded: as oo in *school*
ए	/e:/	Long close-mid front unrounded vowel: as a in *game* (not a diphthong)
ऐ	/æ:/	Long near-open front unrounded vowel: as a in *cat*
ओ	/o:/	Long close-mid back rounded: as o in *tone* (not a diphthong)
औ	/ɔ:/	Open-mid back rounded vowel: as au in *caught*

conditioned allophone of schwa. Thus, the pronunciation of the vowel अ occurs in two forms. When this vowel is followed by word-middle /h/, or it surrounds word-middle /h/, or is followed by word ending /h/, it changes allophonically to short /ɛ/. In all other cases, it is the mid central vowel schwa. Thus, the following words "sehar," "rehna," and "keh" are pronounced as /uɛhɛr/, /rɛhna:/, and /kɛh/ and not as /ʃəhər/, /rəhəna:/, and **/kəh/.** The short-open-back-rounded vowel /ɒ/ (as o in hot) does not exist in Hindi at all, other than for English loanwords. In orthography, a new symbol has been invented for it, i.e., ऑ and included in Hindi phonology. There are some additional vowels traditionally listed in the Hindi alphabet. They are, ऋ (a vowel-like syllabic retroflex approximant), pronounced in modern Hindi as /ri/, used only in Sanskrit loan words. All vowels in Hindi, short or long, can be nasalized except ऑ. In Sanskrit and in some dialects of Hindi (as well as in a few words in Standard Hindi), the vowel ऐ is pronounced as a diphthong /əi̯/ or /ai̯/ rather than as /æ:/. Similarly, the vowel औ is pronounced as the diphthong /əu̯/ or /au̯/ rather than as /ɔ:/. Other than these, Hindi does not have true diphthongs—two vowels might occur sequentially but then they are pronounced as two syllables (a glide might come in between while speaking). The schwa vowel (/ə/) is pronounced very short, otherwise it will be very difficult to pronounce a few words in the absence of schwa. Such a situation arises when there is a consonantal cluster at the end of the word. Thus, for phonological purposes, a word-ending grapheme without a halant or any other vowel diacritic must be treated as consonant ending. The schwa in Hindi is usually dropped (syncopated) in Khariboli even at certain instances in word-middle positions, where the orthography would otherwise dictate so. For example, रुकना (to stay) is normally pronounced as /rukna:/, while according to the orthography, it should have been /rukəna:/ [56].

Hindi possesses a symmetrical ten-vowel system. The vowels [ə], [ɪ], [ʊ] are always short in length, while the vowels [a:, i:, u:, e:, o:, ɛ:,ɔ:] are always considered long. The close vowels, which are considered as the distinction of vowel length in Sanskrit, become the distinction of vowel quality in Hindi. The vowel represented graphically as ऐ (Romanized as *ai*) has been variously transcribed as [ɛ:] or [æ:]. The open central vowel is often transcribed in IPA by either [a:] or [ɑ:]. Hindi is quite similar to English, in contrast to the consonants.

Table 2.2 Consonants of Hindi (borrowed from Persian and Arabic)

Devnagri letters	IPA equivalent
क़	/qə/ voiceless uvular plosive
फ़	/fə/ voiceless labiodental fricative
ख़	/xə/ voiceless velar fricative
ग़	/ʁə/ voiced velar fricative
ज़	/zə/ voiced alveolar fricative
ड़	/ɽə/ unaspirated retroflex flap
ढ़	/ɽʱə/ aspirated retroflex flap

2.3.2   Consonants of Hindi

Hindi traditionally has 28 consonants inherited from earlier Indo-Aryan. Supplementing these are two consonants that are internal developments in specific word-medial contexts [56], and seven consonants originally found in loan words, whose expression is dependent on factors such as status (class, education, etc.) and cultural register (Modern Standard Hindi vs. Urdu).

Most native consonants double in length, except /bʱ, ɽ, ɽʱ, ɦ/. These consonants are known as geminates. Geminate consonants are always medial and preceded by one of the interior vowels such as /ə/, /ɪ/, or /ʊ/. In Hindi consonants, there are four-way distinction of phonation among plosives, whereas, these is the two-way distinction found in English phonology. The phonations are:

tenuis, as /p/, which is like ⟨p⟩ in English spin (with a voice onset time almost zero)
Voiced, as /b/, which is like ⟨b⟩ in English bin
Aspirated, as /pʰ/, which is like ⟨p⟩ in English pin, and
Murmured, as /bʱ/.

The last is commonly called "voiced aspirate." In a study by Shapiro [51], it is demonstrated that the two types of sounds involve two distinct types of voicing and release mechanisms. The series of so-called voice aspirates should be considered to involve the voicing mechanism of murmur in which the air flow passes through an aperture between the arytenoid cartilages, as opposed to passing between the vocal bands. The murmured consonants are a phonation that was lost in all branches of the Indo-European family except Indo-Aryan. In the IPA, the five murmured consonants can also be transcribed as /b̤/, /d̤ /, /ḓ/, /d̤ʒ/ and /g̈/. A few sounds, borrowed from the other languages like Persian and Arabic, are written with a dot (bindu or nukta) as shown in Table 2.2. Many native Hindi speakers, especially those who come from rural backgrounds and do not speak really good Khariboli, pronounce these sounds as the nearest equivalents in Hindi.

ड़ /ɽə/ and ढ़ /ɽʱə/ are not of Persian/Arabic origin, but they are allophonic variants of simple voiced retroflex stops [56]. Most commonly and mainly used vowels and consonants of Hindi phonology are shown in Table 2.3 with their IPA representation.

Table 2.3 Vowels and consonants of Hindi

Vowels	IPA	Consonants	IPA	Consonants	IPA	Consonants	IPA
अ	/ʌ/	क	k	त	t	श	s
आ	/ɑː/	ख	kʰ	थ	tʰ	ष	ʂ
इ	/i/	ग	g	द	d	स	ʃ
ई	/iː/	घ	gʰ	ध	dʰ	ह	ɦ
उ	/u/	ङ	ŋ	न	n		
ऊ	/uː/	च	ʧ	प	p		
ए	/eː/	छ	ʧʰ	फ	pʰ		
ऐ	/æː/	ज	dʒ	ब	b		
ओ	/oː/	झ	dʒʰ	भ	bʰ		
औ	/ɔː/	ञ	ɲ	म	m		
		ट	ʈ	य	j		
		ठ	ʈʰ	र	ɾ		
		ड	ɖ/ɽ	ल	ɭ		
		ढ	ɖʰ/ɽʰ	ळ	l		
		ण	ɳ	व	v		

Chapter 3
Speech Materials and Instrumentation

To a certain extent recording of speech samples and analysis of exemplars depends on the quality and contents of the material prepared for experiments. Therefore, preparing speech materials seems an important part of the research methodology. For this particular case study, various texts have been prepared in Standard Hindi using devnagri script, including the words or phrases commonly found in telephonic interception of criminal acts. Later, these prepared texts were transliterated into each dialect chosen for the study considering the accent as well as the uses.

3.1 Selection of Informants

In order to minimize the intra-dialectal variation due to regions, informants have been selected from a uniform area of the regional dialect. Also they are chosen from the closed age group of 20–25 years with minimum higher secondary education. Condition is that the informant should not have any influence of other native language on his or her dialect. Considering the criteria mentioned above, 15 male and 15 female informants were selected from each dialect. Information related to the informants and their dialectal backgrounds are also noted. Out of the various districts in which a dialect is being spoken, only one place has been chosen to collect the samples of the informants. Name of the regional dialect and the place from where the speech samples of the informants have been collected are mentioned below in Table 3.1.

M. Kulshreshtha and R. Mathur, *Dialect Accent Features for Establishing Speaker Identity:* 21
A Case Study, SpringerBriefs in Electrical and Computer Engineering,
DOI 10.1007/978-1-4614-1138-3_3, © Manisha Kulshreshtha 2012

Table 3.1 Name of the dialect and the place of recording

Name of the regional dialect	Place of recording
Bundeli	Jhansi (Uttar Pradesh)
Khariboli	Anwalkhera (Uttar Pradesh)
Marwari	Alwar (Rajasthan)
Bhojpuri	Deoria (Gorakhpur)
Chattisgarhi	Charoda (Raipur)
Kanauji	Kayanpur (Kanpur)
Haryanvi	Hisar (Haryana)

3.2   Recording of Speech Exemplars

The choice of the microphone depends on the goal of a particular purpose. In this study, the microphone used is a dynamic microphone of Philips (Model DM295) with 1.8 mV/Pa sensitivity, 600 ohms impedance, and frequency range of 100–12,000 Hz.

With the microphone of above-mentioned specifications, speech samples have been recorded directly on computer using inbuilt multimedia card. Speech samples of informants (dialect speakers) are recorded in three repetitions, in Khariboli as well as in their regional dialect.

3.3   Digitization of Speech Samples

Currently, for speech sample analysis, we use computer-based equipment and data is in digital domain. All analog signals need to be converted into digital signals, where Analog-to Digital converter (abbreviated ADC) is used as an electronic circuit that coverts continuous analog signal into discrete digital numbers. Here, various types of ADCs are discussed:

A *successive approximation ADC* uses a comparator and very useful to reject ranges of voltages, eventually settling on a final voltage range. Successive approximation works constantly by comparing the input voltage with a known reference voltage until the best approximation is achieved. In entire process, a binary value of the approximation is stored in a successive approximation register (SAR) and the SAR used as a reference voltage for comparisons. Though this type of ADCs has a good resolution and quite wide range, due to more complex design they are not very popular. A *delta-encoded ADC* has an up-down counter that feeds a digital signal to analog converter (DAC). Both the input signal and the output signal of DAC go to a comparator, which controls the counter. The circuit uses negative feedback from the comparator to adjust the counter until the DAC's output is close enough to the input signal. The number is read from the counter. The great advantage with delta converters is the wide range and high resolution, except the conversion time is dependent on the input signal level.

A *ramp-compare ADC* (also called integrating, dual-slope or multislope ADC) produces a saw-tooth signal in which voltage ramps up and then quickly falls to zero. When the ramp starts, a timer starts counting. When the ramp voltage matches the input, the timer's value is recorded. Timed ramp converters require a least number of transistors. The ramp time is sensitive to temperature because the circuit generating the ramp is often just some simple oscillator. A special advantage of the ramp-compare system is that comparing a second signal just requires another comparator, and another registers to store the voltage value.

A *pipeline ADC* (also called subranging quantizer) uses two or more steps of subranging. In the first step, conversion is done and in the second step the difference to the input signal is determined with a digital-to-analog converter (DAC). This difference is then converted and the results are combined in the last step. This type of ADC is fast, has a high resolution, and only requires a small size.

A *sigma-delta ADC* (also known as a delta-sigma ADC) filters the desired signal band by oversampling of the desired signal by a large factor. ADCs are integral to much current music reproduction technology, since much music production is done on computers; even when analog recording is used, an ADC is still needed to create the PCM (Pulse Code Modulation) data stream that goes onto a compact disk. ADCs are used virtually everywhere where an analog signal has to be processed, stored, or transported in digital form.

3.4 Sampling of Speech Exemplars

The recorded utterances of the informants chosen for the study have been subjected to preliminary auditory analysis for selection of appropriate speech data from the raw data. The utterances are chosen in which the accent features of the speakers are well reflected and are found suitable on the basis of speech quality/clarity of the recorded sample. Speech exemplars of 15 male informants and 15 female informants are chosen in each dialectal group.

3.5 Instrumentation

3.5.1 Microphones

A microphone is a transducer that converts sound energy into electrical energy. Sound information exists as patterns of air pressure and the microphone changes this information into patterns of electric current.

There are a variety of mechanical techniques that can be used to build a microphone. The two most commonly used methods are the magnetodynamic and variable condenser. A majority of microphones used in recording of sounds are either capacitor (electrostatic) or dynamic (electromagnetic) models, which employ a moving diaphragm to capture the sound, but make use of a different electrical principle for converting the mechanical energy into an electrical signal. The efficiency of this conversion is very important because the amount of acoustic energy produced by voices and musical instruments is very small.

3.5.1.1 Dynamic Microphone

In the magnetodynamic, commonly called dynamic microphone, sound waves cause movement of a thin metallic diaphragm and an attached coil of wire. A magnet produces a magnetic field, which surrounds the coil and motion of the coil within this field causes the current to flow. It is important to remember that the current is produced by motion of diaphragm and the amount of current is determined by the speed of that motion. The problem with dynamic microphones is that they are most effective only while working with relatively loud sound sources.

3.5.1.2 Ribbon Microphone

These microphones are comprised of a thin metal ribbon suspended in a magnetic field. When the sound energy is encountered, the electrical signal generated is induced in the ribbon itself. The main advantage of ribbon microphone is its smooth and detailed sound, but this type of microphones is costly and more fragile than conventional dynamic microphones.

3.5.1.3 Capacitor Microphone (Condenser Microphone)

In a condenser microphone, the diaphragm is mounted close to backplate without touching it. The voltage of the battery, the area of the diaphragm, and the backplate and the distance between the two determine the amount of charge. When the distance changes in response to the sound, the current flows in the wire as the battery maintains the correct charge. The amount of the current is essentially proportional to the displacement of the diaphragm and is so small that it must be electrically amplified before it leaves the microphone.

The wide availability of electorate condenser microphones has greatly simplified the problem of obtaining high-quality recordings. Electret microphones respond directly to the sound pressure of the speech signal. Directional electorate microphones respond differentially to sounds coming from one direction. This can be an advantage if one is recording the samples in a noisy environment.

3.5.2 Sound Spectrograph

Alexander Melville Bell in 1867 developed a visual representation of the spoken words later named as "visible speech" at Bell Telephone Laboratory. In 1940s Potter,

Kopp, and Green developed a new method of speech sound analysis using speech spectrograph. Dr. Ralph Potter introduced an electromechanically acoustic spectrograph in 1941. In 1962 Lawrence Kersta, an engineer and a staff member of the Bell laboratories, reexamined the voiceprint method at the request of law enforcement group and introduced the instrument named sound spectrograph as a potential tool for Forensic Speaker Identification. Basic function of this device was to convert the speech into visual representation of its frequency and intensity components.

A sound spectrograph has four parts: a magnetic recorder, electronic filters, a rotating drum on which the spectrogram is recorded, and an electrically operated stylus. The traditional analog version of the sound spectrograph records the input signal on a magnetic medium that goes round the outside edge of the thin drum. Magnetic image is formed on the thin recording disk by the recording head just like a conventional tape recorder when the sound spectrograph switch is on recording mode.

The voice/sound spectrograph is of three types:

Analog spectrograph
Digital spectrograph
Hybrid spectrograph

In *analog spectrograph* speech from the microphone is fed into a band pass filter. Harmonics of the voice whose frequency falls within the range of that filter gives the output with the amplitude proportional to its strength, and then produces a three-dimensional record with the stylus on a paper, showing the change in frequency and amplitude with time.

A *digital spectrograph* consists of special circuits embedded in the microprocessor systems to produce the spectrogram simultaneously with the speech. Voice identification Inc., USA, has produced a real-time digital spectrograph, which produces video display of spectrograms. The spectrograms determine the duration of the speech segments, calculate fundamental frequency and formant ranges, etc.

Hybrid spectrograph is the combination of the two spectrograms mentioned earlier.

The sound spectrograph analysis requires only a limited amount of utterance at a time, which could be recorded in one revolution of the drum. It converts the speech signal into a visual spectrum in the form of traces of the graph. The dimensions and the intensity of the traces are dependent on the utterance being analyzed. Nowadays, the use of computers in conjunction with the spectrograph has increased the volume of the recording.

3.5.3   Computerized Speech Laboratory

Computerized Speech Lab (CSL) for Windows is a hardware and software system for the acquisition, acoustic analysis, display, and playback of speech signals. It records, edits, and quickly analyzes the speech signal. It is possible to carry out the detailed studies of the utterances through segmentation of the recordings.

Fig. 3.1 Computerized Speech Laboratory Model 4300B

CSL is suitable for any acoustic signal characterized by changing spectra over time. It is a Windows-based program, which requires a computer operating under Windows 95 and Windows 98. Operations of CSL include acquisition, storing speech to disk memory, graphical and numerical display of speech parameters, audio output, and signal editing. A variety of analyses, namely, spectrographic analysis, pitch contour analysis, LPC analysis, Cestrum analysis, FFT and Energy contour analysis, etc., can be performed through this instrument. It gives results easily and quickly in comparison with the old speech spectrograph and can handle large speech data at a time. Speech exemplars chosen during the sampling process have been analyzed using Computerized Speech Laboratory model 4300 B as shown in Fig. 3.1.

Chapter 4
Analysis and Results

Speech of person is characterized using some phonetic features like stylistic impression, phonation, nasality, dynamic of loudness, and flow of speech. Based on these features, the linguistic characteristics of a person in specimen are compared with that of the person in questioned. A conventional method, combined aural and spectrographic method, is used for analysis of speech exemplars. Although analysis is performed equally on all the speech exemplars of all chosen regional dialects, due to limited space, Khariboli and Bhojpuri dialects are selected to discuss the results in this chapter. Bhojpuri dialect is selected randomly, whereas Khariboli is considered as a base language in this study. A prosodic analysis is also performed in order to study the intonation tonal pattern of the dialects. This chapter will go through the details of each method used and the results obtained through it.

4.1 Auditory Phonetic Analysis

Phonetic is the study of the phonemes related to a language or dialect and their idiosyncratic pronunciation by a speaker. The technique is based on the process called *critical listening*. Under this technique, a particular speaker is to be identified using phonetic sequences and the phonetic events undertaken while speaking accented speech as well as regional dialect.

In this case study, recorded speech samples were analyzed on the basis of perceptual characteristic. A sentence as well as a word segment was chosen based on the phonetic content and clarity of the utterance. The chosen sentence and word segment are analyzed to study the intonation and tone of speakers belonging to the regional dialects, and the results were compared with the analytical results of Khariboli speakers.

M. Kulshreshtha and R. Mathur, *Dialect Accent Features for Establishing Speaker Identity:* 27
A Case Study, SpringerBriefs in Electrical and Computer Engineering,
DOI 10.1007/978-1-4614-1138-3_4, © Manisha Kulshreshtha 2012

4.2   Vowel Quality and Quantity (Spectrographic Method)

Second step of the identification is spectrographic analysis of the voice samples. In this case study, spectrographic method is used to identify the vowel quality and quantity for the informants of the regional dialect. The vowel quality is a term by which one can identify the difference between two vowel sounds, whereas the length of the vowel is referred as vowel quantity. In order to study vowel quality and quantity of the informants, five different syllable of C_1VC_2 structure are chosen from the recorded utterances for all the dialects. These syllables includes five basic vowels of Hindi, namely, /ə/, /a/, /i/, /o/, and /u/ and are, respectively, represented in chapter as V1, V2, V3, V4, and V5 while illustrating in the graphs. The chosen words and the description of the vowel as nuclei are given in Table 4.1.

4.3   Results of Khariboli and Bhojpuri dialect

During spectrographic analysis, F1 (first formant frequency), F2 (second formant frequency) and duration of syllabic nuclei are measured at appropriate location of the vowel formants while pronouncing one of the selected vowels. Tables 4.2a–4.2e are the tabulation of these measurements for all male subjects of Khariboli while producing vowels /ə/, /a/, /i/, /o/, and /u/, respectively. Similarly, Tables 4.2f–4.2j are the tabulation of F1, F2, and duration measurements for all female subjects of Khariboli while producing vowels /ə/, /a/, /i/, /o/, and /u/, respectively.

Similarly, Tables 4.3a–4.3e are the tabulation of these measurements for all male subjects of Bhojpuri dialect while producing vowels /ə/, /a/, /i/, /o/, and /u/, respectively. Similarly, Tables 4.3f–4.3j are the tabulation of F1, F2, and duration measurements for all female subjects of Bhojpuri while producing vowels /ə/, /a/, /i/, /o/, and /u/, respectively.

These measured values of F1 and F2 are plotted in vowel quadrilateral to obtain the vowel quality with F1 on "y" axis and F2 on "x" axis. Figures 4.1 and 4.2 show the vowel quality reflected by Khariboli male and female speakers for all selected vowels of Hindi. Similarly, Figs. 4.3 and 4.4 show the vowel quality of Bhojpuri male and female speakers. Vowel quality of speakers belonging to other

Table 4.1 Syllables selected for the study

Vowel	/ə/ (V1)	/a/ (V2)	/i/ (V3)	/o/ (V4)	/u/ (V5)
Selected words (C1VC2)	/kəl/	/bat/	/tʰik/	/kon/	/tum/
English meaning	Tomorrow or yesterday	Talk	All right	Who	You
Description of the vowel	Open-mid-back-unrounded	Open-back-unrounded	Close-front-unrounded	Open-mid-back-rounded	Close-back-rounded

Table 4.2a   Values of F1, F2, and duration of Khariboli male informants in vowel /ə/

Speakers	First formant frequency (F1) Hz	Second formant frequency (F2) HZ	Duration of syllabic nuclei (ms)
KhM-1	645	1,392	0.05
KhM-2	646	1,392	0.04
KhM-3	645	1,363	0.05
KhM-4	545	1,349	0.04
KhM-5	674	1,392	0.06
KhM-6	674	1,392	0.04
KhM-7	645	1,363	0.05
KhM-8	674	1,392	0.04
KhM-9	674	1,392	0.05
KhM-10	674	1,349	0.04
KhM-11	645	1,392	0.05
KhM-12	646	1,392	0.04
KhM-13	645	1,363	0.05
KhM-14	545	1,349	0.04
KhM-15	674	1,392	0.06
Average	643	1,378	0.05

Table 4.2b   Values of F1, F2, and duration of Khariboli male informants in vowel /ɑ/

Speakers	First formant frequency (F1) Hz	Second formant frequency (F2) HZ	Duration of syllabic nuclei (ms)
KhM-1	689	1,291	0.14
KhM-2	631	1,291	0.11
KhM-3	674	1,234	0.13
KhM-4	588	1,349	0.14
KhM-5	732	1,306	0.15
KhM-6	689	1,248	0.14
KhM-7	732	1,291	0.13
KhM-8	775	1,234	0.15
KhM-9	714	1,289	0.13
KhM-10	686	1,261	0.11
KhM-11	689	1,291	0.14
KhM-12	631	1,291	0.11
KhM-13	674	1,234	0.13
KhM-14	588	1,349	0.14
KhM-15	732	1,306	0.15
Average	682	1,284	0.13

Table 4.2c   Values of F1, F2, and duration of Khariboli male informants in vowel /i/

Speakers	First formant frequency (F1) Hz	Second formant frequency (F2) HZ	Duration of syllabic nuclei (ms)
KhM-1	430	1,587	0.07
KhM-2	443	1,681	0.03
KhM-3	470	1,573	0.06
KhM-4	430	1,318	0.05
KhM-5	511	1,681	0.04
KhM-6	511	1,762	0.07
KhM-7	484	1,681	0.10
KhM-8	430	1,681	0.06
KhM-9	457	1,587	0.06
KhM-10	457	1,479	0.03
KhM-11	430	1,587	0.07
KhM-12	443	1,681	0.03
KhM-13	470	1,573	0.06
KhM-14	430	1,318	0.05
KhM-15	511	1,681	0.04
Average	460	1,591	0.06

Table 4.2d   Values of F1, F2, and duration of Khariboli male informants in vowel /o/

Speakers	First formant frequency (F1) Hz	Second formant frequency (F2) HZ	Duration of syllabic nuclei (ms)
KhM-1	551	995	0.12
KhM-2	565	982	0.12
KhM-3	590	982	0.12
KhM-4	511	928	0.13
KhM-5	632	1,062	.09
KhM-6	551	928	0.18
KhM-7	618	982	0.14
KhM-8	618	1,062	0.12
KhM-9	551	995	0.14
KhM-10	591	1,008	0.12
KhM-11	551	995	0.12
KhM-12	565	982	0.12
KhM-13	590	982	0.12
KhM-14	511	928	0.13
KhM-15	632	1,062	0.09
Average	575	992	0.13

Table 4.2e   Values of F1, F2, and duration of Khariboli male informants in vowel /u/

Speakers	First formant frequency (F1) Hz	Second formant frequency (F2) HZ	Duration of syllabic nuclei (ms)
KhM-1	551	1,008	0.07
KhM-2	497	1,170	0.06
KhM-3	403	1,035	0.07
KhM-4	390	1,062	0.07
KhM-5	551	1,062	0.07
KhM-6	457	1,143	0.06
KhM-7	565	1,049	0.07
KhM-8	470	1,062	0.07
KhM-9	443	1,116	0.07
KhM-10	497	1,008	0.06
KhM-11	551	1,008	0.07
KhM-12	497	1,170	0.06
KhM-13	403	1,035	0.07
KhM-14	390	1,062	0.07
KhM-15	551	1,062	0.07
Average	481	1,070	0.06

Table 4.2f   Values of F1, F2, and duration of Khariboli female informants in vowel /ə/

Speakers	First formant frequency (F1) Hz	Second formant frequency (F2) HZ	Duration of syllabic nuclei (ms)
KhF-1	738	1,793	0.10
KhF-2	783	1,703	0.08
KhF-3	874	1,778	0.05
KhF-4	768	1,582	0.06
KhF-5	648	1,582	0.05
KhF-6	753	1,718	0.06
KhF-7	814	1,793	0.07
KhF-8	678	1,869	0.04
KhF-9	693	1,703	0.07
KhF-10	783	1,869	0.05
KhF-11	738	1,793	0.10
KhF-12	783	1,703	0.08
KhF-13	874	1,778	0.05
KhF-14	768	1,582	0.06
KhF-15	648	1,582	0.05
Average	756	1,722	0.06

Table 4.2g Values of F1, F2, and duration of Khariboli female informants in vowel /ɑ/

Speakers	First formant frequency (F1) Hz	Second formant frequency (F2) HZ	Duration of syllabic nuclei (ms)
KhF-1	889	1,417	0.11
KhF-2	814	1,417	0.12
KhF-3	949	1,477	0.11
KhF-4	1055	1,417	0.09
KhF-5	904	1,477	0.10
KhF-6	979	1,537	0.10
KhF-7	949	1,703	0.10
KhF-8	919	1,522	0.11
KhF-9	814	1,492	0.11
KhF-10	904	1,613	0.11
KhF-11	889	1,417	0.11
KhF-12	814	1,417	0.12
KhF-13	949	1,477	0.11
KhF-14	1055	1,417	0.09
KhF-15	904	1,477	0.10
Average	919	1,485	0.11

Table 4.2h Values of F1, F2, and duration of Khariboli female informants in vowel /i/

Speakers	First formant frequency (F1) Hz	Second formant frequency (F2) HZ	Duration of syllabic nuclei (ms)
KhF-1	361	2,653	0.10
KhF-2	361	2,683	0.08
KhF-3	361	2,698	0.10
KhF-4	361	2,713	0.12
KhF-5	331	2,502	0.08
KhF-6	407	2,834	0.10
KhF-7	361	2,803	0.11
KhF-8	316	2,502	0.10
KhF-9	407	2,698	0.12
KhF-10	346	2,879	0.10
KhF-11	361	2,653	0.10
KhF-12	361	2,683	0.08
KhF-13	361	2,698	0.10
KhF-14	361	2,713	0.12
KhF-15	331	2,502	0.08
Average	359	2,681	0.10

Table 4.2i Values of F1, F2, and duration of Khariboli female informants in vowel /o/

Speakers	First formant frequency (F1) Hz	Second formant frequency (F2) HZ	Duration of syllabic nuclei (ms)
KhF-1	693	1,160	0.09
KhF-2	663	1,100	0.12
KhF-3	783	1,160	0.11
KhF-4	587	1,010	0.08
KhF-5	919	1,311	0.10
KhF-6	829	1,221	0.05
KhF-7	874	1,447	0.09
KhF-8	783	1,070	0.07
KhF-9	753	1,206	0.07
KhF-10	633	1,010	0.09
KhF-11	693	1,160	0.09
KhF-12	663	1,100	0.12
KhF-13	783	1,160	0.11
KhF-14	587	1,010	0.08
KhF-15	919	1,311	0.10
Average	744	1,162	0.08

Table 4.2j Values of F1, F2, and duration of Khariboli female informants in vowel /u/

Speakers	First formant frequency (F1) Hz	Second formant frequency (F2) HZ	Duration of syllabic nuclei (ms)
KhF-1	542	1,326	0.05
KhF-2	572	1,160	0.09
KhF-3	512	1,281	0.05
KhF-4	437	1,266	0.05
KhF-5	557	1,477	0.04
KhF-6	587	1,477	0.05
KhF-7	587	1,266	0.05
KhF-8	542	1,356	0.04
KhF-9	497	1,582	0.04
KhF-10	587	1,492	0.04
KhF-11	542	1,326	0.05
KhF-12	572	1,160	0.09
KhF-13	512	1,281	0.05
KhF-14	437	1,266	0.05
KhF-15	557	1,477	0.04
Average	536	1,346	0.05

Table 4.3a Values of F1, F2, and duration of Bhojpuri male informants in vowel /ə/

Speakers	First formant frequency (F1) Hz	Second formant frequency (F2) HZ	Duration of syllabic nuclei (ms)
BM-1	646	1,564	0.07
BM-2	612	1,428	0.06
BM-3	578	1,326	0.07
BM-4	578	1,292	0.07
BM-5	612	1,326	0.06
BM-6	714	1,394	0.10
BM-7	612	1,156	0.07
BM-8	578	1,428	0.07
BM-9	680	1,496	0.07
BM-10	578	1,428	0.07
BM-11	680	1,326	0.07
BM-12	476	1,258	0.07
BM-13	680	1,360	0.07
BM-14	646	1,326	0.06
BM-15	612	1,326	0.06
Average	618	1,362	0.07

Table 4.3b Values of F1, F2, and duration of Bhojpuri male informants in vowel /a/

Speakers	First formant frequency (F1) Hz	Second formant frequency (F2) HZ	Duration of syllabic nuclei (ms)
BM-1	748	1,666	0.11
BM-2	646	1,326	0.10
BM-3	680	1,394	0.10
BM-4	646	1,428	0.11
BM-5	816	1,394	0.12
BM-6	782	1,428	0.11
BM-7	680	1,258	0.11
BM-8	714	1,292	0.11
BM-9	714	1,360	0.10
BM-10	748	1,428	0.10
BM-11	680	1,258	0.12
BM-12	612	1,224	0.10
BM-13	782	1,292	0.14
BM-14	748	1,292	0.10
BM-15	646	1,258	0.11
Average	709	1,353	0.11

Table 4.3c   Values of F1, F2, and duration of Bhojpuri male informants in vowel /i/

Speakers	First formant frequency (F1) Hz	Second formant frequency (F2) HZ	Duration of syllabic nuclei (ms)
BM-1	408	2,210	0.04
BM-2	374	1,972	0.06
BM-3	374	2,040	0.05
BM-4	408	1,938	0.04
BM-5	374	2,108	0.04
BM-6	408	2,142	0.09
BM-7	408	2,176	0.11
BM-8	374	2,006	0.07
BM-9	374	2,210	0.09
BM-10	374	2,482	0.04
BM-11	374	1,870	0.06
BM-12	408	2,210	0.05
BM-13	476	1,802	0.08
BM-14	476	2,108	0.06
BM-15	374	2,346	0.06
Average	398	2,108	0.06

Table 4.3d   Values of F1, F2, and duration of Bhojpuri male informants in vowel /o/

Speakers	First formant frequency (F1) Hz	Second formant frequency (F2) HZ	Duration of syllabic nuclei (ms)
BM-1	612	1,224	0.12
BM-2	578	1,088	0.13
BM-3	578	1,020	0.12
BM-4	612	1,020	0.12
BM-5	612	952	0.11
BM-6	612	986	0.13
BM-7	510	952	0.12
BM-8	544	1,054	0.12
BM-9	646	952	0.13
BM-10	578	1,088	0.11
BM-11	578	1,054	0.10
BM-12	476	986	0.11
BM-13	646	986	0.14
BM-14	646	1,156	0.10
BM-15	476	1,054	0.10
Average	580	1,038	0.12

Table 4.3e Values of F1, F2, and duration of Bhojpuri male informants in vowel /u/

Speakers	First formant frequency (F1) Hz	Second formant frequency (F2) HZ	Duration of syllabic nuclei (ms)
BM-1	442	1,360	0.05
BM-2	442	1,224	0.04
BM-3	374	986	0.04
BM-4	510	1,326	0.04
BM-5	374	1,020	0.05
BM-6	442	1,122	0.06
BM-7	510	986	0.08
BM-8	442	1,054	0.06
BM-9	510	1,224	0.05
BM-10	442	1,224	0.04
BM-11	442	1,258	0.03
BM-12	442	1,088	0.07
BM-13	544	1,190	0.07
BM-14	476	1,224	0.05
BM-15	510	1,292	0.05
Average	460	1,171	0.05

Table 4.3f Values of F1, F2, and duration of Bhojpuri female informants in vowel /ə/

Speakers	First formant frequency (F1) Hz	Second formant frequency (F2) HZ	Duration of syllabic nuclei (ms)
BF-1	677	1,779	0.07
BF-2	847	1,610	0.07
BF-3	677	1,779	0.06
BF-4	762	1,779	0.07
BF-5	762	1,610	0.07
BF-6	847	1,864	0.06
BF-7	847	1,694	0.06
BF-8	676	1,694	0.08
BF-9	677	1,610	0.06
BF-10	677	1,440	0.05
BF-11	762	1,525	0.07
BF-12	762	1,694	0.08
BF-13	593	1,440	0.06
BF-14	677	1,694	0.07
BF-15	762	1,610	0.07
Average	734	1,654	0.07

Table 4.3g Values of F1, F2, and duration of Bhojpuri female informants in vowel /ɑ/

Speakers	First formant frequency (F1) Hz	Second formant frequency (F2) HZ	Duration of syllabic nuclei (ms)
BF-1	847	1,610	0.12
BF-2	932	1,610	0.12
BF-3	932	1,779	0.12
BF-4	847	1,525	0.13
BF-5	847	1,440	0.12
BF-6	932	1,694	0.12
BF-7	932	1,610	0.11
BF-8	847	1,610	0.12
BF-9	847	1,694	0.10
BF-10	932	1,525	0.12
BF-11	762	1,440	0.12
BF-12	932	1,610	0.11
BF-13	847	1,525	0.10
BF-14	932	1,525	0.12
BF-15	847	1,525	0.13
Average	881	1,581	0.12

Table 4.3h Values of F1, F2, and duration of Bhojpuri female informants in vowel /i/

Speakers	First formant frequency (F1) Hz	Second formant frequency (F2) HZ	Duration of syllabic nuclei (ms)
BF-1	254	2,711	0.09
BF-2	338	2,796	0.08
BF-3	338	2,711	0.08
BF-4	423	2,711	0.08
BF-5	254	2,627	0.08
BF-6	338	2,796	0.10
BF-7	338	2,711	0.08
BF-8	338	2,627	0.10
BF-9	423	2,457	0.08
BF-10	423	2,711	0.08
BF-11	254	2,711	0.10
BF-12	338	2,881	0.08
BF-13	338	2,542	0.08
BF-14	254	2,542	0.12
BF-15	423	2,542	0.09
Average	338	2,671	0.09

Table 4.3i   Values of F1, F2, and duration of Bhojpuri male informants in vowel /o/

Speakers	First formant frequency (F1) Hz	Second formant frequency (F2) HZ	Duration of syllabic nuclei (ms)
BF-1	762	1,610	0.11
BF-2	762	1,186	0.12
BF-3	762	1,440	0.12
BF-4	677	1,271	0.12
BF-5	762	1,440	0.08
BF-6	762	1,355	0.12
BF-7	932	1,440	0.12
BF-8	677	1,271	0.12
BF-9	762	1,186	0.15
BF-10	762	1,271	0.11
BF-11	762	1,271	0.12
BF-12	932	1,355	0.11
BF-13	762	1,355	0.08
BF-14	847	1,186	0.12
BF-15	762	1,186	0.13
Average	779	1,321	0.12

Table 4.3j   Values of F1, F2, and duration of Bhojpuri male informants in vowel /u/

Speakers	First formant frequency (F1) Hz	Second formant frequency (F2) HZ	Duration of syllabic nuclei (ms)
BF-1	508	1,355	0.08
BF-2	508	1,271	0.08
BF-3	508	1,440	0.07
BF-4	593	1,525	0.08
BF-5	593	1,694	0.06
BF-6	593	1,186	0.09
BF-7	593	1,271	0.08
BF-8	508	1,355	0.11
BF-9	508	1,186	0.08
BF-10	423	1,779	0.08
BF-11	508	1,610	0.07
BF-12	508	1,440	0.06
BF-13	593	1,610	0.05
BF-14	423	1,355	0.08
BF-15	593	1,694	0.06
Average	531	1,451	0.08

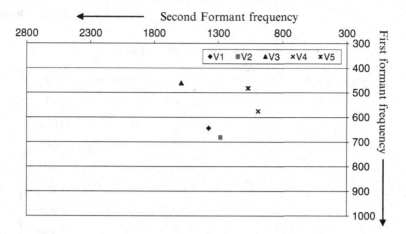

Fig. 4.1 Vowel quadrilateral of average F1 and F2 for Khariboli male informants

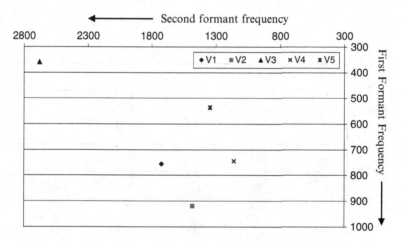

Fig. 4.2 Vowel quadrilateral of average F1 and F2 for Khariboli female informants

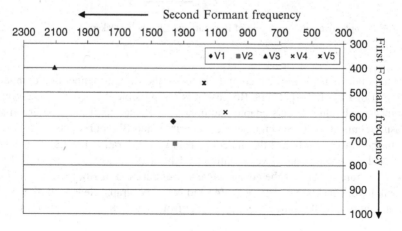

Fig. 4.3 Vowel quadrilateral of average F1 and F2 for Bhojpuri male informants

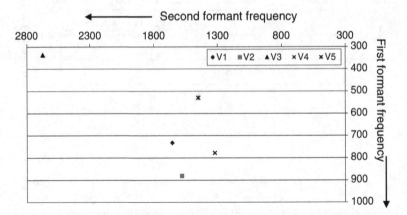

Fig. 4.4 Vowel quadrilateral of average F1 and F2 for Bhojpuri female informants

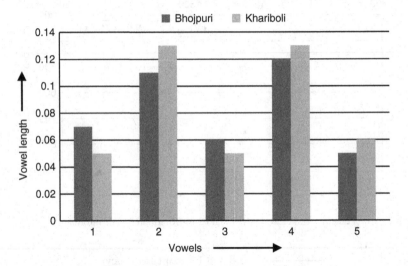

Fig. 4.5 Comparative vowel durations of Bhojpuri and Khariboli speakers

regional dialects is also measured and plotted in the similar manner and compared with that of Khariboli speakers. Because it is not possible to display all the graphs and measured values in this chapter, authors have chosen to display the graph and measurements of the base language, i.e., Khariboli and Bhojpuri dialect. In order to obtain the vowel quantity of the dialects, mean vowel duration for dialect speakers is plotted with the mean vowel duration of Khariboli speakers in case of all five vowels. Figure 4.5 shows the comparative vowel duration of Bhojpuri dialect with Khariboli for vowels V1, V2, V3, V4, and V5. The graph clearly show that the vowel quantity of Bhojpuri dialect is distinct from that of Khariboli speakers.

Figures 4.1–4.4 illustrate the vowel quality of Khariboli and Bhojpuri representing the distribution of formant frequencies, i.e., F1 along Y-axis and F2 along X-axis on vowel quadrilateral. As observed in vowel quadrilateral in Figs. 4.1 and 4.2, 4% of speakers show a tendency of pronouncing a vowel /ə/ more closely than other speakers. Seven percent of speakers show tendency of uttering vowel /ɑ/ more open than others. Twenty percent show tendency to centralize the vowel likewise, and 20% of the speakers also show a tendency of centralizing while pronouncing vowel /i/. The vowel /o/ is pronounced by 13% of speakers more toward vowel /u/, i.e., a closer form of /o/. In the case of vowel /u/, 33% of speakers utter the vowel /u/ a bit open than the other speakers as far as vowel quality distribution is concerned.

From the above observations, central tendency of vowel quality on the speech exemplars of Khariboli male informants for vowel /ə/ is shown by 87% of the speakers, vowel /ɑ/ by 73% of the speakers, vowel /i/ by 80% of the speakers, vowel /o/ by 87% of the speakers, and vowel /u/ by 67% of the speakers. The vowel /i/ is produced like that of /I/ and a bit near to vowel /e/. Central tendencies of the average vowel formant for these vowels are represented in Figs. 4.1 and 4.2 for the utterances of male and female informants, respectively. The vowel /o/ of the informants is between the primary cardinal vowel /O/ and /o/. The vowel /u/ is shown as that of vowel /Y/ and a bit open as well. Interestingly, the vowels of Khariboli male informants show a tendency of producing vowels in the centralized region of the vowel quadrilateral.

Likewise, in case of female informants, vowel quality of the accented Khariboli speakers for vowel /ɑ/ is uttered by 93% of the speakers, vowel /i/ by 80% of the speakers, and vowel /o/ by 80% of the speakers. The vowel quality of vowel /o/ is produced in between the primary cardinal vowel /o/ and /O/ by 80% of the female informants. Vowel quality of vowel /i/ is close to that of vowel /I/ by 80% of the informants.

Figures 4.3 and 4.4 illustrate the vowel quality of Bhojpuri dialect representing the distribution of formant frequencies, i.e., F1 along Y-axis and F2 along X-axis on vowel quadrilateral. Among Bhojpuri male informants, 7% of the male informants produce vowel /ə/ in a bit closer manner than other speakers. Thirteen percent of the speakers produce the vowel more toward the frontal region of the vowel quadrilateral and 20% of the informants produce a mid-open form of vowel /ə/. Seven percent of the male informants produce vowel /ɑ/ more toward the front region of the vowel quadrilateral, 33% of the male informants produce a mid-open form of vowel /ə/, and 7% of the male informants produce a bit closer form of the vowel.

Vowel /i/ is produced toward a bit centralized region of the vowel quadrilateral by 13% of the male informants, 7% of the informants produce a bit open form of vowel, and 13% of the male informants produce the vowel toward the frontal side of the quadrilateral. A bit closer form of vowel /o/ is produced by 20% of the male informants and 13% of the male informants produce the vowel more toward the central region. Thirteen percent of informants produce vowel /u/ in a bit closer manner and 47% of the male informants produce a centralize form of the vowel.

Twenty percent of the female informants produce vowel /ə/ toward the back region of the vowel quadrilateral and 7% of the female informants produce the vowel toward the frontal region. Vowel /ɑ/ is produced toward the back region in a bit closer form by 20% of the female informants, and 7% of the female informants produce the vowel toward the frontal region and a bit opened. Twenty percent of the female informants produce an open form of vowel /i/, and 13% of the female informants produce a frontal form of the vowel. Thirteen percent of the female informants produce an opened form of vowel /o/ and 13% produce the vowel toward the centralized region. Seven percent of the informants produce the vowel /u/ in its opened and centralized forms and 47% of the speakers produce the vowel showing the vowel quality as that of vowel /Y/.

From the above observations, the central tendency of the majority of the Bhojpuri informants for vowel /ə/ is shown by 60% of the male informants, vowel /ɑ/ by 53%, vowel /i/ by 67%, vowel /o/ by 67%, and vowel /u/ by 40%. The vowel quality of vowel /u/ is shown as that of vowel /Y/ and a bit open as well. Likewise, in female informants, vowel quality of the accented Bhojpuri speakers for vowel /ə/ is shown by 73% of the speakers, vowel /ɑ/ by 73% of the speakers, vowel /i/ by 67% of the speakers, vowel /o/ by 67% of the speakers, and vowel /u/ by 47% of the speakers. Central tendencies of the average vowel formant for these vowels are represented in Figs. 4.3 and 4.4 for the utterances of male and female informants, respectively. Forty-seven percent of the female informants produce other form of vowel /o/, i.e., closer form of primary cardinal vowel /o/. The vowel quality of vowel /u/ is shown as that of vowel /Y/ and a bit open as well.

4.4   Prosody Analysis

In general, it is assumed that the pronunciation is what gives an accent to a speech, whereas it is intonation that plays an important role to acquire distinct dialect accent. Therefore, intonation and tone are considered as an essential part of the dialect accent features. Intonation studied on the sentence is called sentence intonation, whereas the intonation observed in a word or word segment is known as lexical tone. Prosody analysis is performed to study the intonation pattern and tonal characteristics of selected regional dialects, yet the results of only Khariboli and Bhojpuri dialects are explained and discussed in this chapter. The intonation pattern of Khariboli male informants is shown in Figs. 4.6a–4.6c, whereas Figs. 4.7a–4.7c represent the intonation pattern reflected by three randomly selected Khariboli female informants along the sentence /kəl//mudʒɛ//bəhʊt//bukʰɑɾ//tʰɑ/. Based on the data, intonation pattern shown by majority of male and female speakers along the sentence is identified to be the intonation pattern of Khariboli dialectal group. Figure 4.8a, b shows the pitch variation along the word **/kɑm/**and hence represents the tone of the Khariboli male speakers Similarly, Fig. 4.8c, d represents the tone of the Khariboli female speakers. On the basis of the analysis for tonal characteristics, a majority of Khariboli speakers use a falling tone. Falling tone is considered as lexical tone in Khariboli accent.

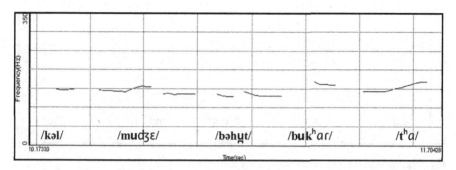

Fig. 4.6a Intonation for the Khariboli speaker KhM-1

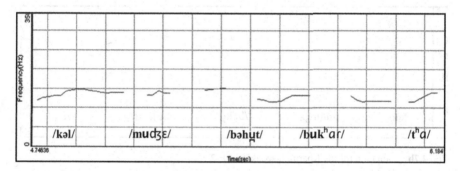

Fig. 4.6b Intonation for the Khariboli speaker KhM-2

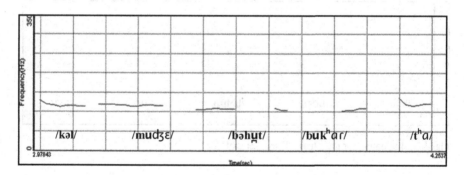

Fig. 4.6c Intonation for the Khariboli speaker KhM-3

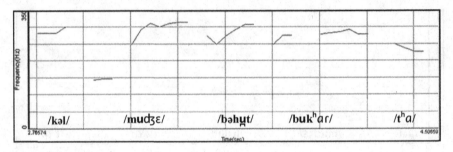

Fig. 4.7a Intonation for the Khariboli speaker KhF-1

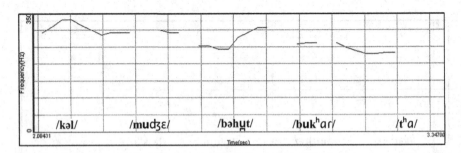

Fig. 4.7b Intonation for the Khariboli speaker KhF-2

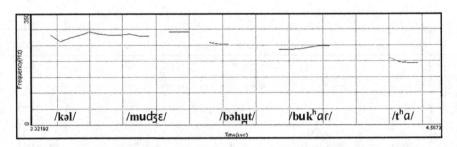

Fig. 4.7c Intonation for the Khariboli speaker KhF-3

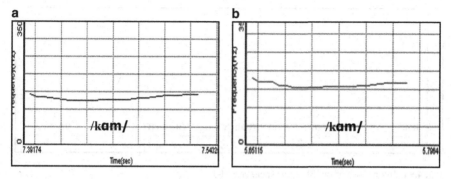

Fig. 4.8a, b Tonal characteristics for the Khariboli male speakers

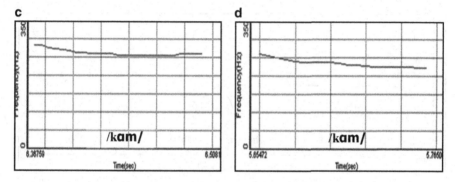

Fig. 4.8c, d Tonal characteristics for the Khariboli female speakers

Now, for Bhojpuri dialect, a prosody analysis is performed on the selected sentence and word. Sentence intonation and lexical tone has been observed and represented in terms of raising or falling pitch pattern. Figures 4.9a–4.9c show the intonation pattern of three randomly selected Bhojpuri male informant, whereas Figs. 4.10a–4.10c show the intonation pattern for three randomly selected female informant. Tonal pattern of Bhojpuri male informants is represented in Fig. 4.11a, b and for female informants it is shown in Fig. 4.11c, d.

Similar results were obtained for each subject of each dialect for males as well as for females. Due to space limitation, it is not possible to explain and display the intonation and tone for all the dialect groups in this chapter. The specification of CSL instrument at the time of analysis and measurement was as follows: sampling rate of 25 KHz, which is down sampled at 12.5 KHz and wide band analytical filter of 183.11 Hz, has been used with the quantization at 16 bits.

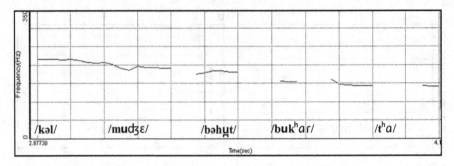

Fig. 4.9a Intonation for the Bhojpuri speaker BM-1

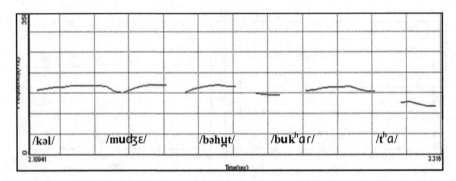

Fig. 4.9b Intonation for the Bhojpuri speaker BM-2

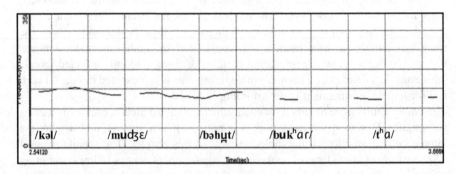

Fig. 4.9c Intonation for the Bhojpuri speaker BM-3

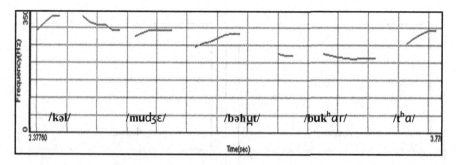

Fig. 4.10a   Intonation for the Bhojpuri speaker BF-1

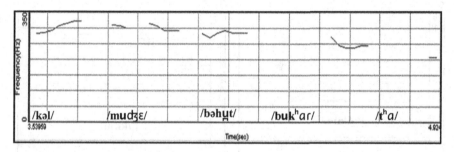

Fig. 4.10b   Intonation for the Bhojpuri speaker BF-2

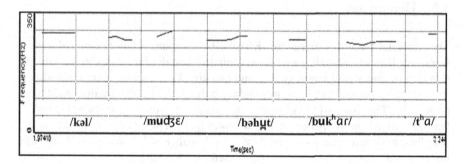

Fig. 4.10c   Intonation for the Bhojpuri speaker BF-3

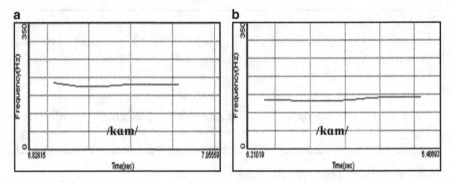

Fig. 4.11a, b Tonal characteristics for the Bhojpuri male speakers

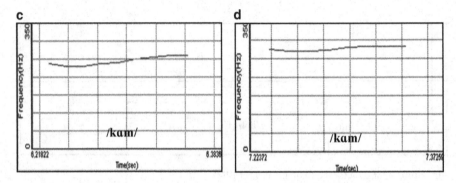

Fig. 4.11c, d Tonal characteristics for the Bhojpuri female speakers

4.5 Discussion

The dialectal accent features of Khariboli, Bhojpuri, Chattisgarhi, Marwari, Kannauji, Bundeli, and Haryanvi have similarity and dissimilarities in the intonation pattern, tones, and vowel quality and quantity. In order to explain the findings from auditory-phonetic and spectrographic analysis, the results are discussed here profoundly. The observed vowel quality and quantity on vowel quadrilateral revealed that 60% of Bhojpuri speakers show a vowel quality in a closer form than 87% Khariboli male informants for vowel /ə/. Vowel quality of /ɑ/ used by 53% of male informants is a closer form of the vowel used by 73% Khariboli male informants. In case of vowel /i/, a majority of the Bhojpuri speakers use a frontal form of the vowel than that of a majority of Khariboli male informants. On the other hand, vowels /o/ and /u/ are a closed form of the vowels in case of Bhojpuri speakers than that of the native speakers of Khariboli. The observations in female informants are in agreement with the observations in the utterances of male informants. Interestingly, female Bhojpuri speakers show more consistent results than male informants.

As far as vowel quantity or vowel length is concerned, the average vowel length of Bhojpuri speakers is more than the vowel length of Khariboli speakers for vowels /ə/, /ɑ/, and /i/, whereas for vowels /o/ and /u/ Bhojpuri speakers use short vowel length than Khariboli speakers. Overall, vowel quality of a majority of Bhojpuri dialect speakers is different from Khariboli speakers for chosen vowels as the native speakers of Bhojpuri dialect use a bit closer form of the vowels than the vowels used by Khariboli speakers. Vowel quality of Chattisgarhi dialect speakers is different from Bhojpuri for vowel /ə/, which is used in an opened form and vowel /i/ is pronounced very close to the cardinal point by the majority of Chattisgarhi speakers. Most of the Marwari speakers have a distinguishable vowel quality in case of vowel /u/ than Bhojpuri and Chattisgarhi speakers. Marwari speakers have a vowel quality more toward the back region of the vowel quadrilateral for vowel /u/. Kannauji speakers have a distinctive feature for vowel /o/, which is used in a more open manner than other dialects speakers, where they have a tendency to shift the vowels quality toward the central region of the vowel quadrilateral. Vowel quality of Bundeli is similar to Bhojpuri except vowel /u/, which is used in open manner by Bundeli speakers. Haryanvi speakers pronounce vowel /o/ in a more open form, which is similar to Kannauji dialect but vowel /i/ is used very close to the cardinal point of the vowel in case of Haryanvi dialect as well. If we look at the vowel quantity, it is perceived that the average vowel quantity of Bhojpuri and Bundeli speakers is more or less similar to that of Khariboli speakers. Vowel quantity of Kannauji speakers is similar to the average vowel quantity of Marwari speakers in comparison with Khariboli.

The sentence prosody and lexical prosody of the speakers belonging to the chosen dialects are found to be dialect specific and carries the accent information of the individual. The findings are supported by the study conducted by Cummins and reported that lexical and sentence prosodic features for identification of accented feature of a dialectal group or language and lexical tone and phrasal intonation is an important marker of accent and stress in every language [21]. In terms of acoustic information, which is related to prosody, vowel quality, and vowel quantity, nonnative speakers can be distinguished from native speakers. The finding is in agreement with the study conducted on nonnative speakers of Hindi by native speakers of Punjabi, Dogri, and Kashmiri [51].

Based on the analysis of intonation pattern of Khariboli speakers, a generalized result can be drawn from the initial portion of the sentence and from the end portion of the sentence. The informants show two types of intonation at the initial utterance of the sentence, i.e., a rising type of intonation (H) and a falling type of intonation (L). 53.3% of male informants use rising intonation at the initial utterance of the sentence, whereas 46.7% informants use falling intonation. The observation among the female informants is of interest in which 70% of them use falling intonation and 30% of them show rising–falling pattern of intonation at the starting of the sentence. As far as intonation pattern at the ending of the sentence is concerned, 70% of male informants end with rising pattern and 30% with a falling pattern, whereas 90% of female informants use a falling pattern at the end of the sentence, and only 10% shows a rising pattern. From the above observation, intonation pattern shown by 70% of the male informants at the initial of the sentence and 90% of the female informants at the end is identified as the intonation pattern of Khariboli dialectal group. On the basis of the analysis for tonal characteristics on chosen words, 40%, 30%, and 30% of male informants show level tone, falling tone, and falling–rising tone, respectively. On the other hand, 70% of female informants use a falling tone and 20% and 10% of speakers use level and falling–rising tone, respectively. Tonal characteristics shown by 70% of the female informants with falling tone is considered as lexical tone in Khariboli accent.

In Bhojpuri dialect, a rising intonation pattern is observed in 80% of the male informants at the starting portion of the sentence and 20% of the male informants show falling intonation. On the other hand, 86% of the female informants show rising intonation at the initial of the sentence and rest of the speakers show the falling pattern. Similarly, 67% of the male informants show falling intonation and 33% of informants show rising intonation pattern at the end of the sentence. Likewise, 73% of female informants show falling intonation and rest of the female informants show rising pattern. The individual characteristics are shown at the onset and end of the sentence in which some of the speakers start with low rising, whereas some speakers start with low falling intonation. Similarly some of the speakers end with low falling intonation and some with high falling intonation. However, the intonation pattern at the subsequent part of the onset in the sentence is showing common features among the Bhojpuri speakers. Intonation pattern shown by the majority of the informants (67% of male informants and 73% of female informants) at the subsequent part of the sentence is identified as a Prescription Model for Bhojpuri native speakers.

Figures 4.11a, b and Fig. 4.11c, d illustrate the tonal characteristics of Bhojpuri male and female informants, respectively. As observed in the figures, 40% of the male informants show falling tone in the chosen word, 40% show rising, and 20% of the male informants show level tone. Likewise, 40% of the female informants show falling–rising tone, 27% show rising, 20% show falling, and 13% of the female informants show level tone on the chosen word.

On the basis of the tonal characteristics shown by the Bhojpuri male and female informants, it has been observed that the lexical prosody in case of Bhojpuri dialect cannot be ascertained, as the common features cannot be identified.

The intonation pattern of Bhojpuri speakers at the end of the chosen sentence is observed to be more or less similar to that of the Khariboli informants. Bhojpuri dialect speakers initiated with a falling intonation pattern and Khariboli initiated with a rising intonation pattern. In the case of lexical prosody, the tonal feature of Bhojpuri dialect could not be ascertained; however, Khariboli speakers show a lexical prosody of falling tone.

The discussed results notably indicate that the regional dialects of Hindi are distinct based on the vowel quality, quantity, and prosody information. This could potentially be very useful in forensic speaker identification practice, where the questioned speech samples and specimen are in different language/dialects. In addition, the study will be useful for investigators to narrow down the list of suspects by speaker profiling.

Chapter 5
Conclusion

This case study contributes to understand the speaker identification process, in a situation, where the unknown speech sample is in different language/dialect from the recording of suspect. Hindi, being the most spoken language in India, is selected for the study and its seven popular dialects are chosen including Khariboli as the base language. Speech samples of 20 male and 20 females from each regional dialect are recorded and analyzed to obtain the distinctive features. These features, when combined together, are found useful in the area of forensic voice identification.

The spectrographic study of vowel quality and quantity for various dialects reveals that each dialect possesses its own vowel quality, which is distinguishable while, compared with Khariboli and also the vowel quantity is quite distinct when compared with each other. Therefore, vowel quality is proved to be a useful feature for profiling of speakers, yet there are a few similarities among the regional dialect speakers. The features, based on the vowel quality, are cooperated with features of prosody as discussed later in the chapter.

Vowel quantity analysis reflects various observations; speakers of Bhojpuri, Chattisgarhi, Kannauji, Marwari, Bundeli, and Haryanvi dialects use longer vowel /ə/ as compared with Khariboli. Speakers of Bhojpuri, Chattisgarhi, Bundeli, Kannauji, and Marwari dialects use longer vowel /a/ as compared with Khariboli, whereas speakers of Haryanvi dialect use short vowel. Dialectal speakers of Chattisgarhi, Kannauji, Haryanvi, and Marwari dialects use longer vowel /i/ as compared with Khariboli, whereas speakers of Bhojpuri and Bundeli dialects use short vowel. Likewise, speakers of Bhojpuri, Chattisgarhi, Kannauji, Bundeli, and Marwari dialects use long vowel /o/ as compared with Khariboli, whereas speakers of Haryanvi dialectuse short vowel. Dialectal speakers of Chattisgarhi and Haryanvi dialects use long vowel /u/ as compared with Khariboli, whereas speakers of Bhojpuri, Kannauji, Bundeli and Marwari dialects use short vowels.

This study also suggested that specific acoustic features of Bhojpuri, Chattisgarhi, Kannauji, Marwari, Khariboli, Bundeli, and Haryanvi dialects, based on the results of prosody analysis, are unique and distinguishable for profiling of speakers belonging

M. Kulshreshtha and R. Mathur, *Dialect Accent Features for Establishing Speaker Identity: A Case Study*, SpringerBriefs in Electrical and Computer Engineering, DOI 10.1007/978-1-4614-1138-3_5, © Manisha Kulshreshtha 2012

to the regional dialectal group, validating that the acoustic features associated with lexical and sentence intonation are useful for speaker profiling.

Quality of front vowels' expressed in terms of first and second formants is found more significant as profile characteristic than that of back vowels. Thus, overall distribution of vowel quality on vowel quadrilateral is very critical in characterization of dialectal accent based speaker profiling. Due to exposure to other nonnative dialects, there is a chance of variation in accent features, although accent features are likely to remain as profile characteristic for quite some time. Some of the speakers' vowel quality and quantity is deviating from the vowel quality and quantity of the accented dialect. The observations imply that some of the regional dialect speakers use vowel quality and quantity other than the vowel quality and quantity of their own dialect, maybe because of the influence of other language or dialect. It is very important that words with similar vowel quality and quantity should be chosen as clue word. In actual crime case examination, if the auditory impression is of accented utterances in either questioned or specimen speech exemplars, it is recommended to study the vowel quality and quantity of the utterances in the questioned sample as well as in the specimen sample. Consequently, an enormous number of possible syllabic nuclei of same vowels is required to be selected and the preliminary study of variant of vowels within questioned as well as within specimen is required to be conducted in order to understand the variant of vowel quality due to the production of accented and unaccented utterances by the accused or by the suspect(s).

As the population distribution of speakers of Kannauji is in close proximity to the population distribution of speakers of Khariboli, there are similarities in features among the native speakers of Khariboli and Kannauji dialects. Although Haryanvi dialect speakers' distribution is in close proximity to the Khariboli dialect speakers' distribution, there are larger distinctive features among the native speakers of Haryanvi and Khariboli as the Haryanvi dialect of Hindi is heavily influenced by the Panjabi on the northwest of Haryanvi region. Likewise, the Bundeli and Chattisgarhi dialect speakers are showing some similarities in the initial and the end of the sentence intonation. However, differences occur in the middle of the sentence. The clear distinction between Bundeli and Chattisgarhi can be made in terms of the vowel quality, where Bundeli speakers are showing a closer form of the vowels than the Chattisgarhi speakers. The study reveals that most of the female native speakers of the regional dialect are showing a constraint to the process of language change as far as accented delivery is concerned. The identification is better among female speakers than the male speakers. Most of the dialect speakers of Hindi are found to be different in terms of tonal characteristics when compared with Khariboli. In this study, a majority of the speakers of the chosen dialects use tonal features different from the tonal features of Khariboli speakers.

However, there is influence of regional dialects or deviated accented features among the Khariboli speakers, identification with larger number of representative data, i.e., $(80\pm10\%)$ is possible. Likewise, Haryanvi speakers can also be identified with the larger representative data. Other dialects of Hindi, namely, Bhojpuri, Chattisgarhi, Kannauji, Marwari, Khariboli, and Bundeli are observed to be influenced by nonnative accent and lesser number of representative data for identification is used specially among the male speakers. The observations imply that there is a constant influence of other regional dialect on the native accented features of these dialects of Hindi. Dialect accent of male speakers are found more variable than the female speakers. The discussed findings encourage us to collect similar data from other dialects of Hindi in order to create a database for speaker profiling. Such database is one of the very important and informative databases for forensic labs working with Speaker Identification.

References

1. Abberton E, Fourcin AJ (1978) Intonation and speaker identification. Lang Speech 21(4):305–318
2. Ahmed R, Agrawal SS (1969) Significant features in the perception of Hindi consonants. J Acoust Soc Am 45(3):758–763
3. Alexander LF, Valter C (2001) Lexical tone contrast effects related to Linguistic experience. J Acoust Soc Am 109(5):2475
4. Allard J, Wayland R, Wong S (2000) Acoustic characteristics of English fricatives. J Acoust Soc Am 108(3):1252–1263
5. Atal BS (1974) Effectiveness of linear prediction characteristics of speech wave for automatic speaker identification and verification. J Acoust Soc Am 55(6):1304–1312
6. Atkinson JE (1976) Inter and intra speaker variability in fundamental voice frequency. J Acoust Soc Am 60(2):440–445
7. Black J, Lashbrook W, Nash E, Oyer H, Pedry C, Tosi O, Truby H (1973) Reply to speaker identification by speech spectrograms: some further observations. J Acoust Soc Am 54: 535–537
8. Bolt RH, Cooper FS, David EE, Denes PB, Picket JM, Stevens KN (1970) Speaker identification by speech spectrograms: a scientist view of its reliability for legal purpose. J Acoust Soc Am 47:597–612
9. Bolt R, Cooper F, David E, Denes P, Picket J, Stevens K (1973) Speaker identification by speech spectrograms: some further observations. J Acoust Soc Am 54:531–534
10. Braun B, Kochanski G, Grabe E, Rosner BS (2006) Evidence for attractors in English intonation. J Acoust Soc Am 119(6):40006–4015
11. Bricker P, Pruzansky S (1966) Effect of stimulus content and duration on talker identification. J Acoust Soc Am 40(6-II):1441–1449
12. Caroline RW, James DH (2006) The influence of Gujrati and Tamil L1s on Indian English: a preliminary study. Word Englishes 25(1):91–104
13. Chao YR (1933) Tone and intonation in Chinese. Bull Inst Hist Philol 4:121–134
14. Cheung RS, Eisenstein BA (1978) Feature selection via dynamic programming for text independent speaker identification. IEEE Trans Acoust Speech Signal Process ASSP-26(5): 396–403
15. Chiba T, Kajiyama M (1941) The vowel: its nature and structure. Kaiseikan, Tokyo
16. Clarke FR, Bricker RW (1969) Comparison techniques for discriminating among talkers. Speech Hearing Res 12:747–761
17. Coleman R (1973) Speaker identification in the absence of inter subject differences in glottal source characteristics. J Acoust Soc Am 53:1741–1743
18. Connell B (2000) The perception of lexical tone in Mambila. Lang Speech 43(2):163–182

19. Cooper WE, Serensen JM (1981) Fundamental frequency in sentence production. Springer, New York
20. Crystal D (1969) Prosodic systems and intonation in English. Cambridge studies in linguistics. Cambridge University Press, Cambridge
21. Cummins F, Gers F, Schmidhuber J (1999) Comparing prosody across many languages. I.D.S.I.A. Technical Report, IDSIA-07
22. Dara C, Pell MD (2006) The interaction of linguistic and affective prosody in a tone language. J Acoust Soc Am 119(5):3303–3304
23. Das SK (1969) A method of decision making in pattern recognition. IEEE Trans18:329–333
24. Dhirendra Verma (1996) Hindi Bhasha aur Lipi. Hindustani Academy, Allahabad
25. Dik JH, Joost VG (1991) The frequency scale of speech intonation. J Acoust Soc Am 90(1):97–102
26. Donald JS, Hemeyer T (1972) Identification of place of consonant articulation from vowel formant transitions. J Acoust Soc Am 51(2):652–658
27. Fry DB (1970) Prosodic phenomenon. In: Malmberg B (ed) Manual of phonetics. North Holland, Amsterdam
28. Fry DB (1979) The physics of speech. Cambridge University Press, Cambridge
29. Fujimura O (1962) Analysis of nasal consonants. J Acoust Soc Am 34:1865–75
30. Fujimura O (1971) Sweep tone measurements of vocal characteristics. J Acoust Soc Am 49(2):541–558
31. Glenn JW, Norbert K (1968) Speaker identification based on nasal phonation. J Acoust Soc Am 43(2):368–372
32. Gray CH, Kopp GA (1944) Voice print identification. Bell Telephone Laboratories Report, New Jersey, pp 13–14
33. Green N (1972) Automatic speaker recognition using pitch measurements in conversational speech. Joint Speech Research Unit, Report No. 1000
34. Grierson GA (1928) Linguistic Survey of India. Vol I-XI, Calcutta, ISBN 81-85395-27-6
35. Hazen B (1973) Effects of differing phonetic context on spectrographic speaker identification. J Acoust Soc Am 54(3):650–658
36. Hillenbrand J, Getty LA, Clark MJ, Wheeler K (1995) Acoustic characteristics of American English vowels. J Acoust Soc Am 97:3099–3111
37. James DH (1996) Pitch range and focus in Hindi. J Acoust Soc Am 99(4):2493–2500
38. Kenneth NS (1966) Acoustical description of syllabic nuclei: an interpretation in terms of a dynamic model of articulation. J Acoust Soc Am 40(1):123–131
39. Kersta LG (1962) Voice print identification infallibility. J Acoust Soc Am 34(12):1978–1978
40. Koenig EB (1986) Spectrographic voice identification: a forensic survey. J Acoust Soc Am 79(6):2088–2090
41. Ladefoged P (1962) Elements of acoustic phonetics. University of Chicago Press, Chicago
42. Ladefoged P (2001) A course in phonetics, 4th edn. Harcourt, Fort Worth, p 177
43. Lori HH, Andrew J, lotto k, Kluender R (2001) Influence of fundamental frequency on stop-consonant voicing perception: a case of learned co-variation or auditory enhancement. J Acoust Soc Am 109(2):764–774
44. Nolan F (1983) The phonetic bases of speaker recognition. Cambridge University Press, Cambridge
45. Ohala JJ (1978) Production of tone. In: Fromkin VA (ed) Tone: a linguistic survey. Academic, London, pp 5–39
46. Ohala JJ (1983) Cross-language use of pitch: an ethological view. Phonetica 40:1–18
47. Peterson GE, Lehiste I (1960) Duration of syllable nuclei in English. J Acoust Soc Am 32:693–703
48. Rose P (2002) Forensic speaker identification. Forensic science series. Taylor and Francis, London
49. Ruth HM, And Elaine RS (2003) Hemisphere differences in prosody production: a new look. Int Soc Phonet Sci 87:9–17

50. Ryo K, Youngon C (1999) Effects of native language on the perception of American English /R/ and /L/: a comparison between Korean and Japanese. ICPh 99, San Francisco, pp 1429–1432
51. Shapiro MC (2003) "Hindi". In: Cardona G, Jain D (eds) The Indo-Aryan languages. Routledge, New York, pp 250–285, ISBN 9780415772945
52. Singh CP, Singh SR (1998) Voice spectrographic study of class characteristics of Hindi utterances of Punjabi, Dogri and Kashmiri speakers. J Indian Acad Forensic Sci 37(1&2):40–45
53. Smrkovski L (1975) Study of speaker identification by aural and visual examination of non contemporary speech samples. Jr Off Anal Chem 59:927–937
54. Stevens SS, Volkman J (1940) The relation of pitch to frequency: a revised scale. Am J Psychol 53:329–353
55. Stevens KN, Williams CE, Carbonell JR, Woods B (1972) Speaker authentication and identification. A comparison of spectrographic and auditory presentations of speech material. J Acoust Soc Am 51:2030–2043
56. Tiwari B ([1966] 2004) हिन्दी भाषा (Hindī Bhāshā). Kitāb Mahal, Allahabad, ISBN 81-225-0017-X
57. Tosi O, Oyer M, Lashbrock W, Pedey C, Nical J, Nash E (1972) Experiment on voice identification. J Acoust Soc Am 51:2030–2043
58. Web Page /http://www.phonam.de/sprecher_e.html/
59. Wolf JJ (1972) Efficient acoustic parameters for speaker recognition. J Acoust Soc Am 51(6):2044–2057
60. Woo N (1969) Prosody and phonology. Ph.D. dissertation, MIT
61. Yi Xu, Wallace A (2004) Multiple effects of consonant manner of articulation and intonation type on F0 in English (A). J Acoust Soc Am 115(5):2397
62. Young M, Campbell R (1967) Effect of contexts on talker identification. J Acoust Soc Am 42:1250–1254

Index

M. Kulshreshtha and R. Mathur, *Dialect Accent Features for Establishing Speaker Identity:*   59
A Case Study, SpringerBriefs in Electrical and Computer Engineering,
DOI 10.1007/978-1-4614-1138-3, © Manisha Kulshreshtha 2012